Endocrine Control of Skeletal Maturation

Ze'ev Hochberg

Endocrine Control of Skeletal Maturation

Annotation to Bone Age Readings

70 figures and 2 tables, 2002

KARGER

Basel · Freiburg · Paris · London · New York · New Delhi · Bangkok · Singapore · Tokyo · Sydney

Ze'ev Hochberg, MD, DSc
Division of Pediatric Endocrinology
Rambam Medical Center
POB 9602
Haifa 31096 (Israel)

Library of Congress Cataloging-in-Publication Data

Hochberg, Z.
Endocrine control of skeletal maturation:
annotation to bone age readings / Ze'ev Hochberg.
p. ; cm.
Includes bibliographical references and index.
ISBN 3 – 8055 – 7313 – 8 (hbk. : alk. paper)
1. Children – Age determination. 2. Skeletal maturity. 3. Hand – Radiography.
4. Wrist – Radiography. 5. Bone densitometry.
6. Bones – Metabolism – Endocrine aspects.
I. Title.
[DNLM: 1. Age Determination by Skeleton. 2. Bone Development – physiology.
3. Bone Diseases. WE 200 H685b 2002]
RJ132 .H634 2002
612.7'5 – dc21 2002038707

www.karger.com
Printed in Switzerland on acid-free paper by Reinhardt Druck, Basel
ISBN 3 – 8055 – 7313 – 8

To Naomi

Bone of my bones
(Genesis, 2:33)

Contents

Preface

Students and practitioners of child growth have long expressed a need for a small book that clearly delineates the relationship between theory and practice of fundamental bone biology as manifested in routine X-ray films. The basic concept that inspired the purpose of this book was that by holding an X-ray film of the hand and wrist for a reading of 'bone age', the clinician is afforded a unique glimpse at the growing and maturing bones of a child. For me, as a clinician and scientist studying child growth, the growth plate is an exciting place to look at, provided that one has an understanding of the biological basis of the growth process. In the book, I have tried as much as possible to correlate the radiological appearance of normal and abnormal bones with the histology and pathology of the relevant condition. I hope that learning this text will inspire clinicians to read the skeletal maturity films themselves in order to understand the biological basis and control mechanisms of bone growth.

My personal conceptions of bone maturation and growth are based on my own experience. Yet, the preparation of a book such as this inevitably owes great debt to many researchers who have investigated bone growth and maturation. Their names are too numerous to list here, although many are detailed in the reference index. It has not been found feasible to give detailed citations of the vast literature. One name that cannot go unmentioned is that of a bone pathologist, the late Henry L. Jaffe, who, as I learnt, utilized a similar approach to the one presented here over 30 years ago, collecting specimens and figures from his predecessors. Some of the pictures in this book were prepared from this historical collection. Many colleagues granted me permission to use figures from their clinical files and scientific work; their names are cited in the 'Photographic credits' and I am thankful to them all.

A special debt of gratitude goes to Dina Lewinson and Gila Maor from the Department of Anatomy at the Technion Faculty of Medicine, Haifa, for teaching me the fundamentals of bone and cartilage morphology. An old Hebrew idiom says: 'I have learnt understanding from all my teachers, but more so from my pupils.' In that respect, I would like to extend personal thanks to Martin Ritzen, who invited me to give a series of lectures in Stockholm, during which the concept of this book crystallized.

Ze'ev Hochberg, Haifa

Introduction

In the dialectics of human anthropology and auxology, 'bone age' is felt to be an expression of the biological maturity of a child. Inferring from bone maturity, the clinician can contemplate diagnostics and evaluate height prediction, as well as give his/her recommendations for physical activity or advise on the timing for orthodontic procedures and orthopedic surgery. Moreover, pediatricians and endocrinologists feel that bone maturity assessment by hand and wrist radiography gives a reflection of the child's biological age. This is accomplished in a variety of methods; but all of them compare a given film to various standards either averaging or summarizing the maturity of several bones followed by the designation of a 'bone age'. Whereas the aim is noble, the means are naive. A radiogram of the hand and wrist can at best reflect the maturity of the bones that are depicted on that particular film. Even in terms of bone maturation, the recognition of bones' shapes and changes in configuration provide relatively little insight into the maturational processes, all the more so as one utilizes the method of averaging or summarizing the results of several bones to attain a unified reading for the entire film.

More importantly, the hand and wrist film may provide information that is not being evaluated when reading for a 'bone age'. Although the radiogram only reveals calcified elements, it provides a glimpse into a whole variety of processes in bone and cartilage growth, differentiation and calcification – factors which, in turn, are regulated by control mechanisms. Thyroid hormones, sex steroids, calcium-regulating hormones and growth hormones (GH) are but a few of these control mechanisms, and their effect may become evident on careful examination of the hand and wrist radiograms. Comprehension of these mechanisms may infer an underlying endocrine action or derangement and thus lead to a more meaningful reading of the radiogram.

The anatomical and histological maturation of long bones and cuboid bones as represented in the hand and wrist radiogram are reviewed here, and the endocrine regulation of these maturational processes and attempts to uncover endocrine function and malfunctions as they unfold in the radiogram are summarized briefly. The presently used methods of skeletal maturity assessment are reiterated and some doubts about the entire paradigm and its uncertainties are raised, suggesting that beyond simply reading for 'bone age', the hand and wrist films may be used in understanding skeletal maturity in a more meaningful way (**figure 1**).

Following early reflections and the pioneer work of Greulich and Pyle [1], other methods have been developed for the assessment of skeletal maturity. They all average or summarize several bones of the hand and wrist to define a 'bone age'. The very act of averaging entails a loss in distinction of individual segments. The following pages will attempt to show the importance of individual segments in evaluating the mechanisms and pathophysiology of bone growth and maturation.

The terminology 'bone age' and the assignment of 'years' as its units are meaningless. Age, expressed in years, is a precise measurement with no variance and in our age and aging-conscious civilization it entails the meaning of a calendar age. On the other hand, 'bone age' is neither; its reading is imprecise, children at a certain chronological age often have a wide variation in their skeletal maturity and children with different calendar ages have similar 'bone ages'. This book will reason against that abuse of terms. The units for skeletal maturity have accordingly been 'years' in inverted commas. This term is often misleading for children and parents who perceive a retarded or advanced 'bone age' as a physical impairment. It also misleads medical professionals to apprehend it as a means of evaluating the child's biological age. The entry in Medline does not even read 'bone age' and is erroneously labeled 'age determination by skeleton'. This misconception rises to its peak when a court reassigns a child's calendar age according to a 'bone age' film. Would it not make more sense to use the child's mental or psychomotor or developmental age for that purpose? I underscore this misconception and suggest that we omit the 'age' in this term and speak of skeletal maturity or skeletal development and also suggest that skeletal maturity be assigned units of 'maturity index' rather then 'years'. We can still use the available figures from the standards of references. I recommend saying 'This child, at the chronological age of 10 years, has a skeletal maturity index of 9'.

Fig. 1
The 'bone maturity' radiogram **(a)** provides a glimpse into the mechanisms of bone and cartilage growth, differentiation, ossification and calcification **(b)**.

1a

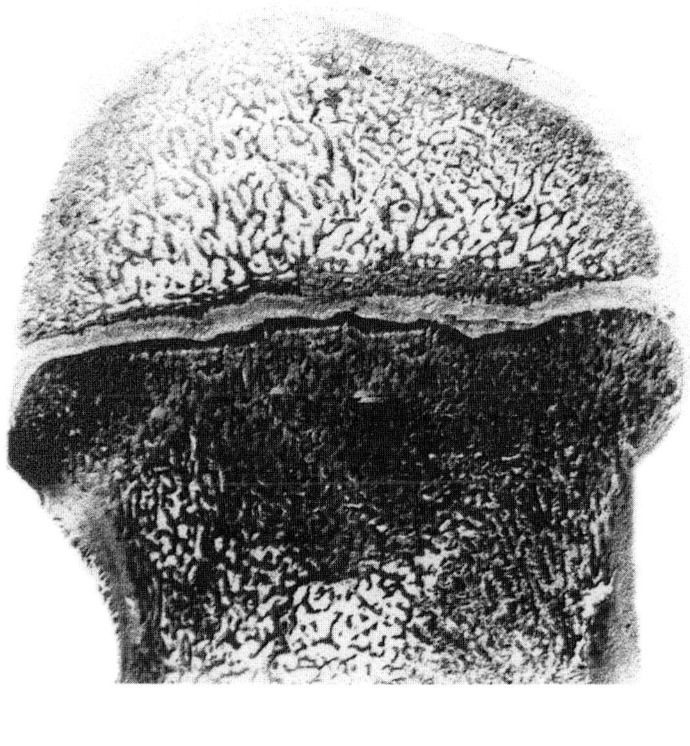

1b

The progression (or tempo) of skeletal maturity correlates weakly with puberty, chronological age or growth velocity [2]. These poor correlations reach their extreme in cases of complete dissociation of skeletal maturity and growth, which would ultimately bring about the eunuch habitus. The tempo of bone maturity does correlate better, however, with the degree of ponderosity. This implies that the unique metabolic and endocrine milieu of obesity regulates this independent index, and as such it needs to be viewed as a measure of this milieu.

It is noteworthy that in the insightful earliest attempt to develop skeletal maturity standards of reference, as early as 1937 Todd [3] called for an Atlas of Skeletal Maturation, emphasizing the maturation process and not the age equivalence. So many years ago, he sensibly made the comment that '... differences in age equivalents are not the result of *variations of the normal* but are sequels to subclinical modifications of time relationships of the several successive features of the maturation process'. The methods developed by Tanner and Whitehouse start similarly with maturation scores, but then change direction into the 'bone age' paradigm.

The concept of a maturity index rather then a 'bone age' also does away with the need for population-specific values or with regular updates due to the secular trend. In analogy to the staging mode of puberty, bone maturity index, as a measure of several control mechanisms, is prone to genetic and environmentally-derived variations.

To understand this relationship I shall briefly review the physiology of bone maturation and its endocrine regulation.

Bone Maturation and Growth

Primary and Secondary Ossification Centers

In the young embryo the skeleton develops out of the primordial hyaline cartilage. **Figure 2** shows the succession of events from a hyaline cartilage, as it transforms into a primary and then a secondary ossification center. Nutrient vessels deliver pluripotent stem cells, which give rise to the bone marrow as a source of building blocks for the ossification centers. Only after the primary and secondary ossification centers grow concentrically and verge upon each other does a growth plate ensue to become the site of enchondral ossification.

Figure 3 shows photomicrographs of the cartilage in the hands of 7-, 9- and 10-week-old embryos [4]. Even before the mesenchymal anlage of the bone transforms centrally into pre-cartilage, the eventual joints are well delineated. Chondrocytes mature and hypertrophy at the site where the primary ossification center would later evolve. Within the cartilage nodule, a bone forms and expands gradually to replace the entire cartilage.

In the case of long bones, the central part of the shaft (diaphysis) is the embryonic site of the primary ossification center **(figure 4)**. Later in the course of maturation, secondary ossification centers would develop at either or both ends of the bone (epiphysis). As primary and secondary ossification centers approach each other, a growth plate ensues. In the case of cuboid bones, such as carpal bones of the hand or vertebrae, primary ossification occurs later in infancy or childhood in the bone center **(figure 5)**. As hyaline cartilage transforms to become a nodule of growth cartilage, groups of young chondroblasts proliferate several times before the cell cycle rests. These cells then hypertrophy at the expense of their matrix in the center of the cartilage nodule. Hypertrophic cells generate a proteoglycan-rich matrix that, on the one hand, allows calcium deposits to surround the cells and, on the other hand, blood vessels and later on bone marrow to penetrate the matrix, mobilizing osteoprogenitor cells to invade the site of hypertrophy and calcification. Now the stage is ripe in the calcified cartilage for programmed death of the hypertrophic cartilage cells and invasion of the remaining lacunae by bone-resorbing and bone-forming cells. Bone is deposited on residues of calcified cartilage.

Fig. 2
Bone develops from its mesenchymal anlage. Hyaline cartilage transforms into a primary, then a secondary ossification center. Nutrient vessels deliver pluripotent stem cells, as a source for building blocks for the ossification centers. Between the primary and secondary ossification centers, a growth plate becomes the site of enchondral ossification.

Fig. 3
Cartilage models of hands.
a Photomicrograph presenting an early stage in the delineation of two contiguous short tubular bones in the hand of a 7-week embryo. The mesenchyme in the interior of these rudiments has already been transformed into precartilage. The eventual joints are well delineated. × 75.
b Photomicrograph of hand bones of 7- and 9-week-old fetuses. Capsules already close the joints. In the mid-portions the cartilage cells are hypertrophic (see **c**). × 8.2.
c Photomicrograph of the proximal phalanx of the middle finger in **b**. The chondrocytes mature and hypertrophy. In the very center cartilage cell lacunae start to form. A thin layer of osseous tissue has formed under the perichondrium (periost). × 40.
d Photomicrograph showing the midportion of a cartilage model of the metatarsus of a 10-week-old fetus. Within the cartilage nodules, hypertrophic cells have secreted calcified matrix. It will appear and expand gradually to replace the entire cartilage by bone. × 92.

Fig. 4
Primary and secondary ossification centers. The central part of the shaft is the embryonic site of the primary ossification center. Secondary ossification centers develop at one or both ends of the bone (epiphysis). As primary and secondary ossification centers approach each other, a growth plate evolves.

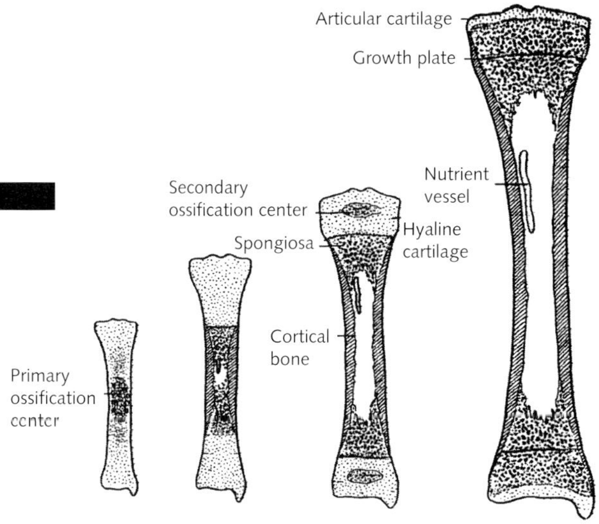
Articular cartilage
Growth plate
Nutrient
vessel
Secondary
ossification center
Spongiosa
Hyaline
cartilage
Cortical
bone
Primary
ossification
center

2

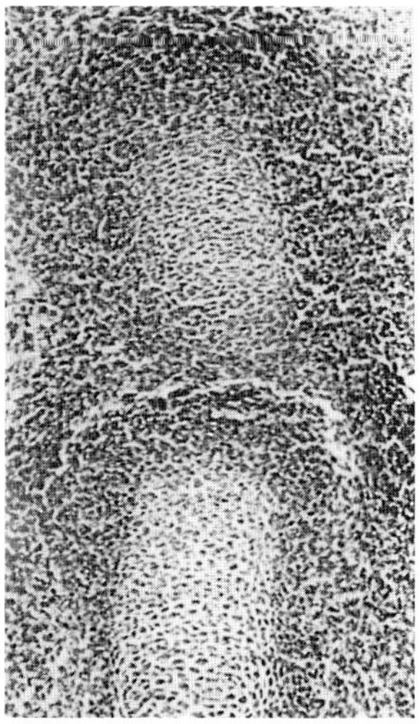

3a

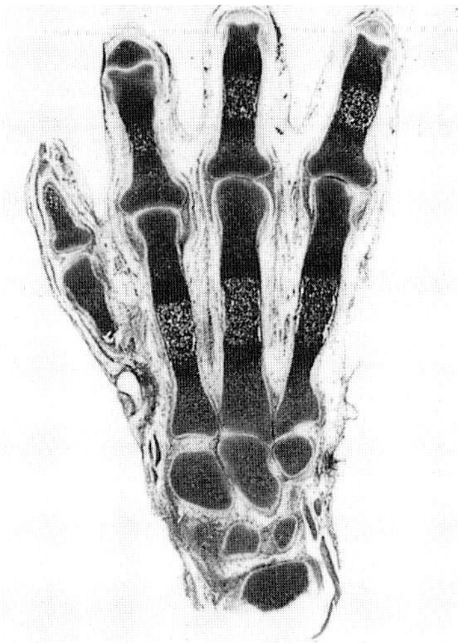

3b

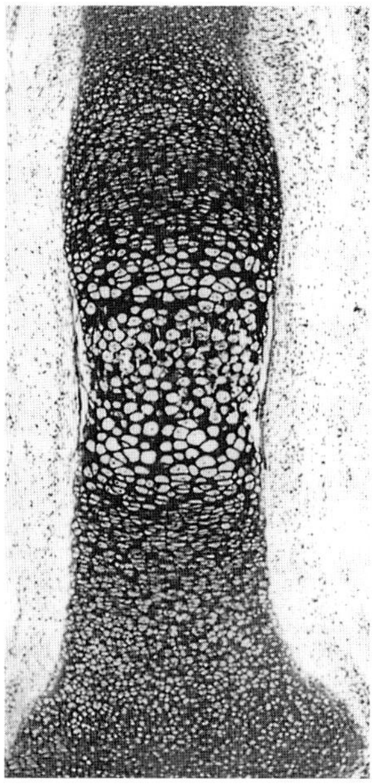

3c

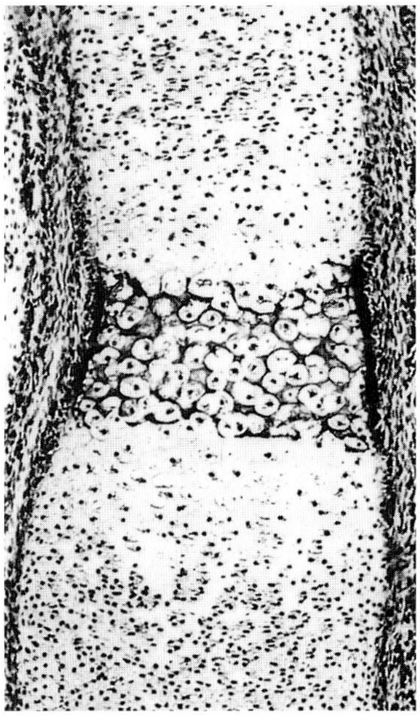

3d

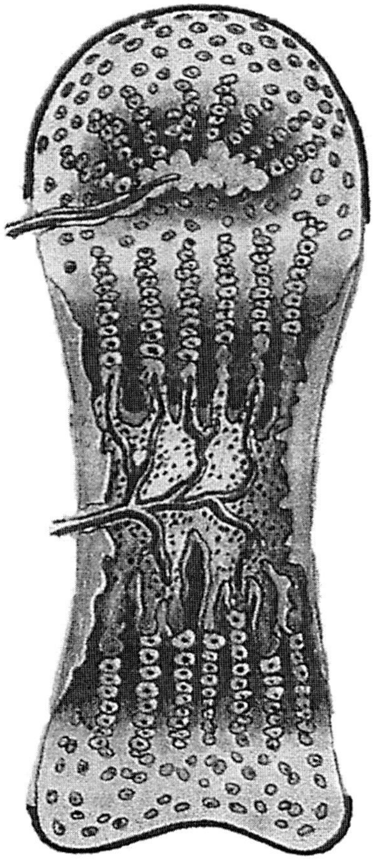

4

Secondary ossification centers develop at one or both ends of the epiphysis **(figure 5)**. The process is similar to that of the primary ossification center. As two centers of a long bone expand and approach each other, a growth plate evolves between them. Chondrocytes at either ends rearrange in cartilaginous nodules, proliferate and the same sequence of events of proliferation, central hypertrophy, calcification and degeneration that established the first primary ossification center is repeated.

In the short bones of the hand, metacarpals and phalanges, two nodules of chondro-epiphyses initially form, but only one end forms a secondary ossification center and only this one end would contribute to the longitudinal growth of the bone. At the other end, cartilage would progressively be replaced by bone from the primary ossification center and finally retains only a thin articular cartilage. In clinical conditions, when growth is hyperstimulated, accessory epiphyses may emerge in otherwise unusual ends of a bone. **Figure 6** shows such accessories after initiation of thyroxine replacement therapy in a hypothyroid child resulting in rapid catch-up growth.

Fig. 5
Primary ossification center in a cuboid bone.
a Schematic presentation. A nourishing vessel develops into the future bone marrow that provides the machinery to transform cartilage cells into bone.
b Photomicrograph illustrating a sagittal section through a cuboid bone. The primary ossification center stands out clearly with the same components as illustrated in **a**. × 39.

Fig. 6
Accessory ossification centers. After initiation of thyroid replacement therapy in this 10-year-old hypothyroid girl, accessory unusual secondary ossification centers developed at the proximal ends of the 2nd to the 5th metacarpals and in the distal end of the 1st metacarpal.

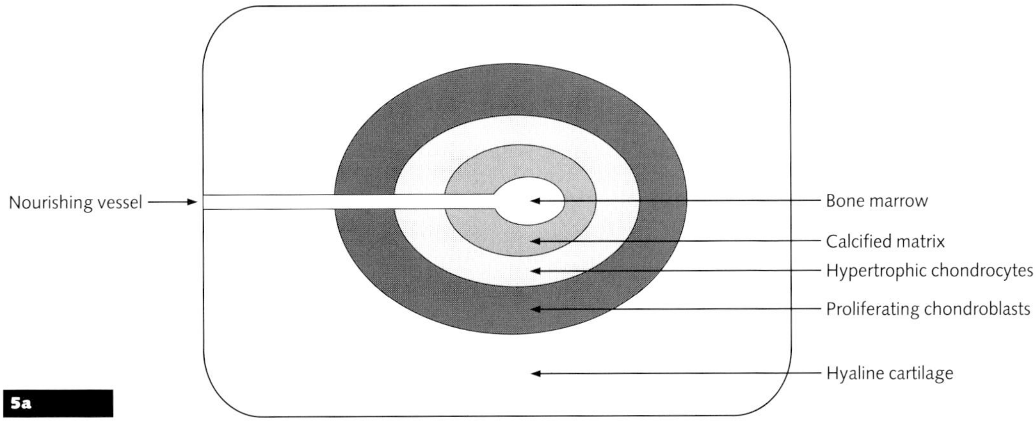
Nourishing vessel
Bone marrow
Calcified matrix
Hypertrophic chondrocytes
Proliferating chondroblasts
Hyaline cartilage

5a

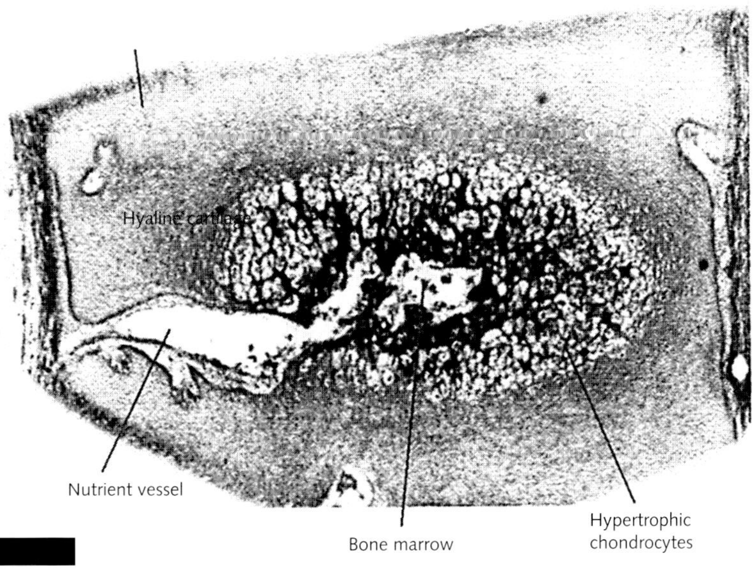
Hyaline cartilage
Nutrient vessel
Bone marrow
Hypertrophic chondrocytes

5b

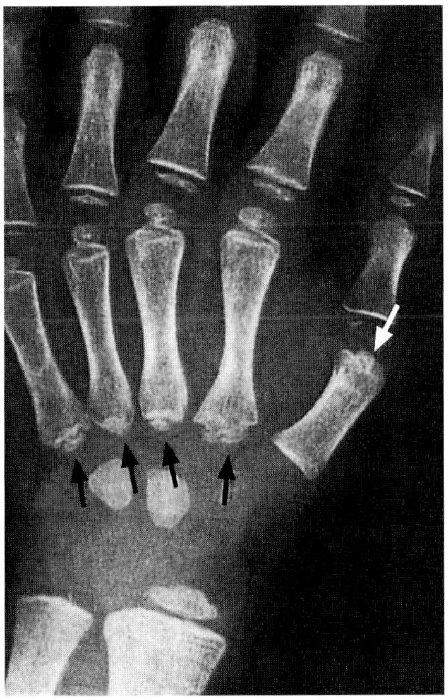

6

Longitudinal Growth at the Growth Plate

Positioned between the epiphyseal bone crest and its metaphyseal boundary, the growth plate is a highly organized functional unit, commissioned to ensure continuous delivery of new cartilage cells and matrix leading to successive bone formation **(figure 7)**. This process, termed enchondral ossification, can be divided for didactic purposes into two consecutive stages:

(1) Chondroplasia – the process of cartilage differentiation and growth.

(2) Osteogenesis – the process of calcification and ossification of cartilage tissue.

This classification is instructive as these two processes can be read separately from a hand and wrist radiograph and because each of them is regulated by its own particular mechanism.

Fig. 7
Enchondral ossification.
a Schematic model, showing the relative contribution of each stage to longitudinal growth. For didactic purposes it is divided into *chondroplasia* – the process of cartilage growth and differentiation and *osteogenesis* – the process of calcification and ossification of the cartilage tissue.
b Photomicrograph of longitudinal cell columns of the growth plate. Adjoining the upper epiphysis, the first layer of reserve cells is followed by proliferating chondroblasts, hypertrophic chondrocytes and the resorption front (RF). Metaphyseal trabecular bone (MTB) is shown in the subchondral region.

Chondroplasia

In the first layer of cells, adjacent to the epiphysis, chondroprogenitor cells form the germinative 'reserve cells', also known as 'resting cells'. This mesenchymal anlage does not express cartilage-specific antigens, but is committed to differentiate into chondrocytes. As part of the cartilage anlage, this zone is vascularized [6] and thus provides a vanguard for hormonal action. One of these hormones is, for example, GH. As detailed later, GH plays a major role in the function of reserve cells and their differentiation into cartilage. Following a step of differentiation, these cells become young cartilage cells, proliferating chondroblasts which undergo rapid divisions along the bone axis, producing longitudinal cell columns **(figure 7)**. Following a limited number of some four cell cycles, the chondrocyte's cell cycle rests, it ceases to divide and starts a phase of cell hypertrophy, along with synthesis of matrix macromolecules. It has been estimated that chondrocyte proliferation contributes to 15% of longitudinal growth and that accumulated extracellular matrix contributes to another 25%.

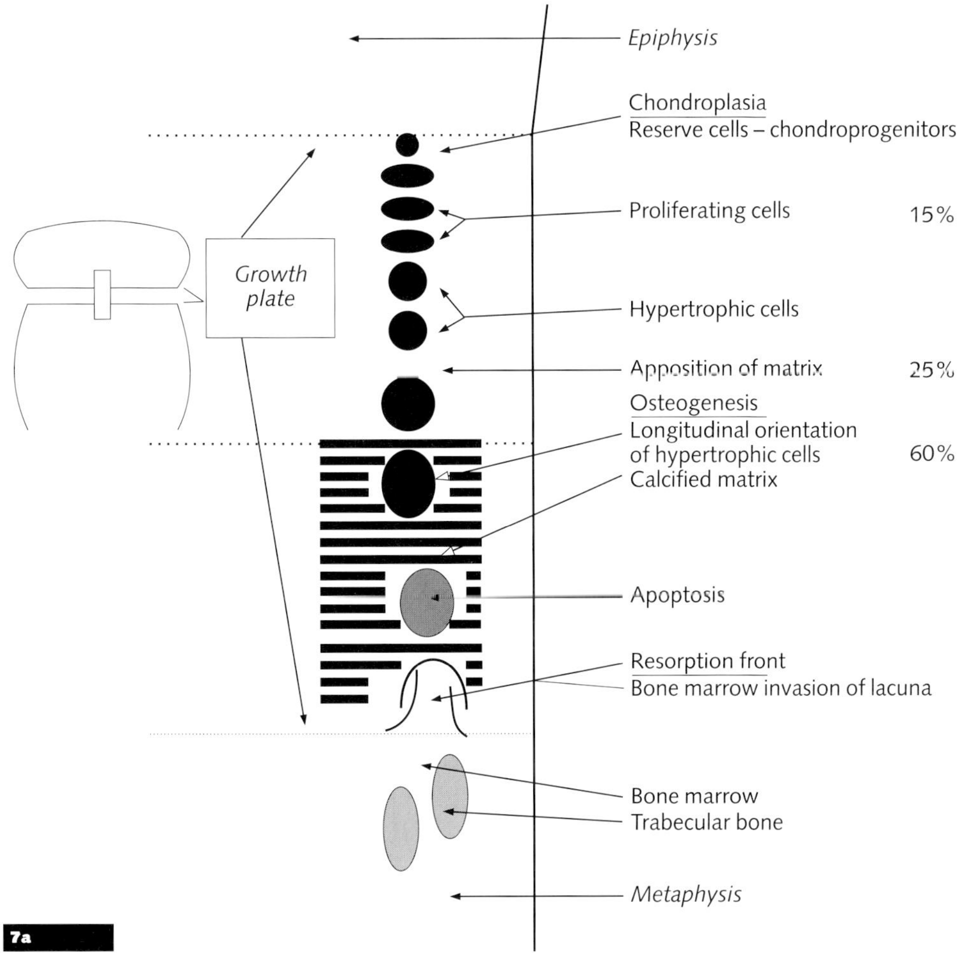

Contribution to growth
Epiphysis
Chondroplasia
Reserve cells – chondroprogenitors
Growth plate
Proliferating cells
15%
Hypertrophic cells
Apposition of matrix
25%
Osteogenesis
Longitudinal orientation of hypertrophic cells
60%
Calcified matrix
Apoptosis
Resorption front
Bone marrow invasion of lacuna
Bone marrow
Trabecular bone
Metaphysis

7a

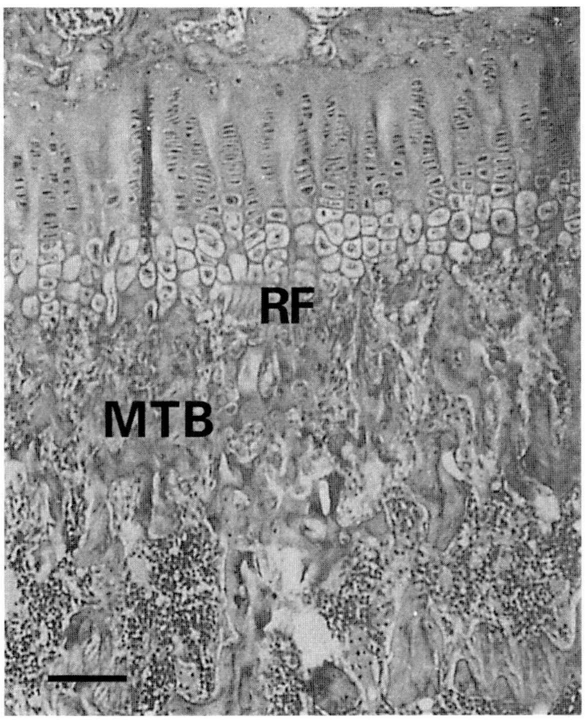

RF
MTB

7b

Within a period of 2 days, the cell changes its orientation from perpendicular to longitudinal. The cell volume increases 4- to 10-fold through a cellular, water-driven swelling mechanism and matrix volume increases 2- to 3-fold. But most importantly, cell height (along the bone axis) increases more than 5-folds, from 7 μm of the flat proliferating chondroblast to 38 μm of the last hypertrophic chondrocyte [7, 8] **(figure 8)** [9].

It was estimated that the relative contribution of cell elongation at the hypertrophic zone to bone growth amounts to 44–60% of the total elongation at the growth plate, depending on the specific bone [10]. A dense calcified matrix further characterizes the distal part of the hypertrophic cell zone. The terminal layers of hypertrophic cells generate submicroscopic deposits of calcium phosphate which are similar to hydroxyapatite and thus, the extracellular matrix of that region undergoes 'cartilage mineralization' **(figure 9)**. Calcified matrix around hypertrophic cells is most evident in the intercellular space within a column then at the intercolumnar matrix. Entrapped in this matrix, hypertrophic cells undergo programmed death (apoptosis), leaving a lacuna that awaits to be invaded by bone marrow and bone-forming elements **(figure 10)**. Hypertrophic chondrocytes express angiogenic protein vascular endothelial growth factor (VEGF). This protein is an essential coordinator for chondrocyte death, chondroclast function and extracellular matrix remodeling, but it also attracts marrow invasion and influences bone formation [11]. Thus, death of the hypertrophic cell is a regulated process which is essential for transforming cartilage into bone tissue [12].

Fig. 8
Proliferating and hypertrophic chondrocytes. Electron micrograph of a rat proximal tibia. Compared to a longitudinal height of 7 μm of the flat proliferating cell **(a)**, the hypertrophic height is 38 μm **(b)**, serving as the most prominent mean of longitudinal bone growth. IM = Intercolumnar matrix; CM = calcified matrix. × 2,000. Scale bar = 4 μm.

Fig. 9
Calcified collagen matrix. Photomicrograph of the proximal end of a human tibia showing the epiphyseal growth plate between the epiphysis and metaphysis. Calcified matrix around hypertrophic cells is more prominent in the intercellular space within a column than it is at the inter-columnar matrix.

Fig. 10
Bone marrow invasion. Photomicrograph of the lower hypertrophic zone. Entrapped in mineralized matrix, hypertrophic cells undergo programmed death, leaving a lacuna that is invaded by bone marrow and bone-forming elements. × 460.

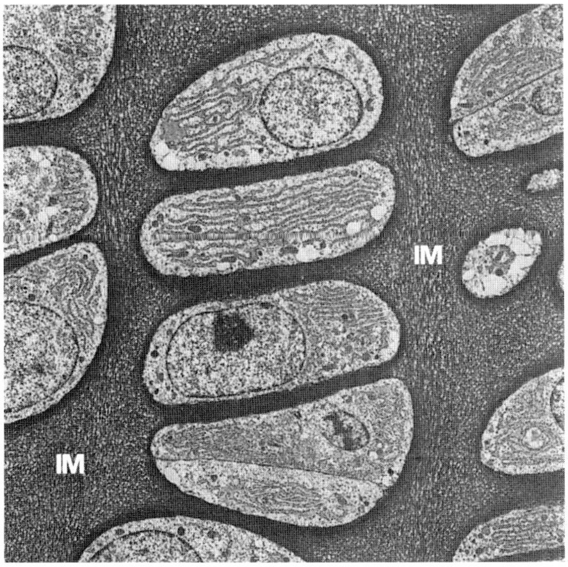

8a

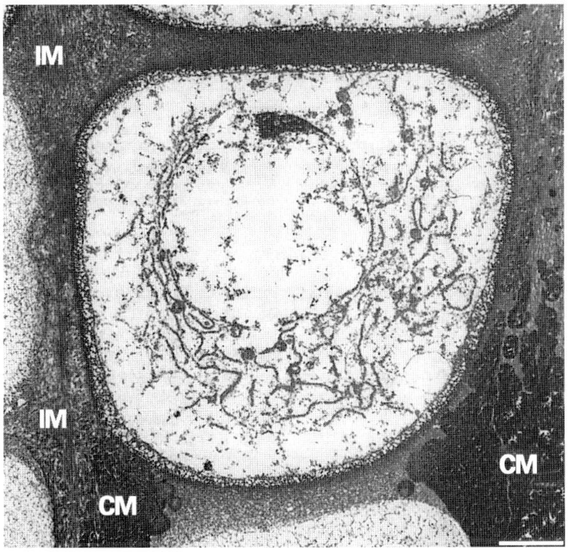

8b

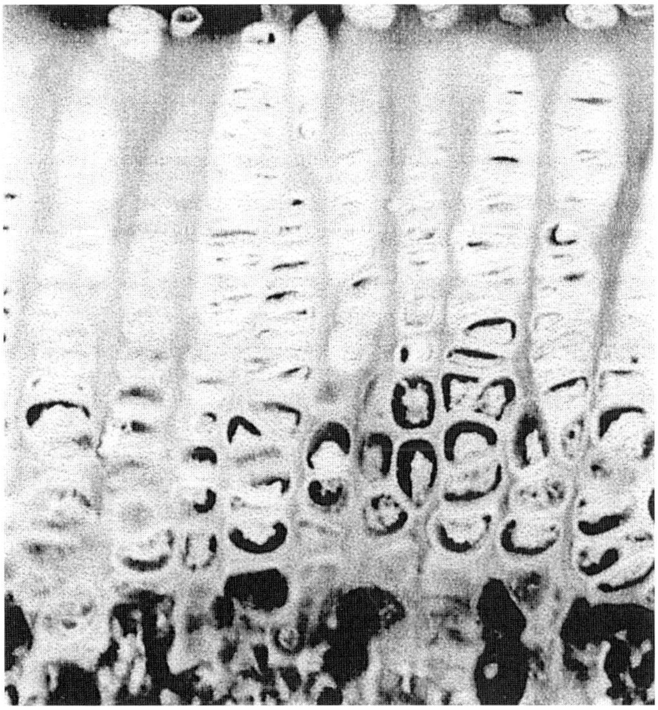

9

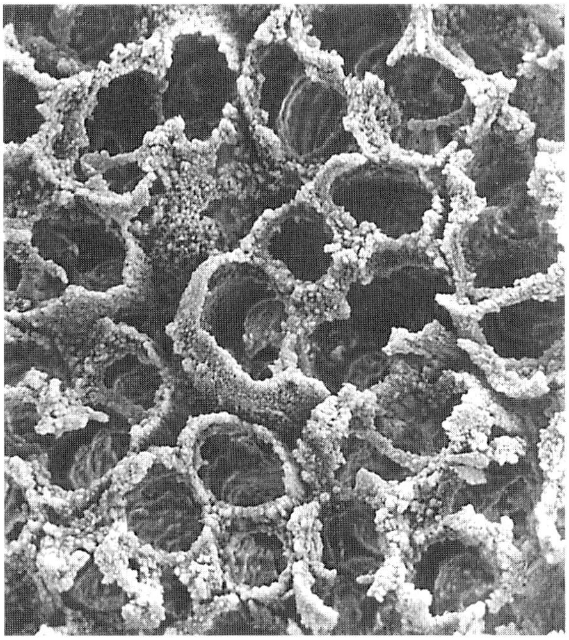

10

Several steps of osteogenesis are mediated by chondrocyte apoptosis. These include intracellular calcium accumulation and release, matrix calcification, matrix resorption, maintenance of growth plate homeostasis, attraction of blood vessels and bone marrow with their osteoblast precursors, and stimulation of bone formation [13]. Apoptosis of terminal hypertrophic cells results in the formation of a lacuna in the calcified matrix **(figure 11)** [14], which further expands through water absorption by matrix hyaluronan [15], and is then invaded by the adjacent bone marrow. Invasion of the calcified cartilage by bone marrow follows through erosion of the collagen tissue by chondroclasts/osteoclasts. The latter cells produce matrix-resorbing proteinases [16] and end in a ruffled border to give rise to penetration of cellular elements and paracrine growth factors from the marrow. Monocytes are fused to become such resorbing osteoclasts, which then initiate the bone remodeling cycle of resorption-formation, modulated by marrow cells and their growth factors **(figure 12)** [17]. The smooth function of the cartilage-bone interface is the last step in osteogenesis. It is the first phase of bone modeling and was believed to be the impaired mechanism in patients with osteogenesis imperfecta **(figure 13)**.

Fig. 11
Mineralizing front of the proximal growth plate cartilage of rat femur, showing calcified lacunae. Scanning electron microscopy of the inorganic specimen. The unmineralized zones have been digested. Mineralized matrix surrounds the (unseen) hypertrophic chondrocytes and in their last layer represents lacunae or apoptosed cells.

Fig. 12
Resorption front osteoclasts. Sagital section through rat proximal tibiae, reacted for tartrate-resistant acid phosphatase as a marker of osteoclasts of the resorption front (RF). SP = Spongiosa. The thick arrows indicate large multinucleated osteoclasts and the thin arrows show mononuclear osteoclasts or osteoclast precursors. Bars = 150 μm.
a Normal rat growth plate (GP).
b GH-deficient rat.

Fig. 13
Photomicrograph of the distal tibia growth plate of a full-term neonate with osteogenesis imperfecta who died 2 days after birth. Chondroplasia is normal and so is matrix calcification, with extensions into the metaphysis. But only a meager amount of osseous tissue is deposited on these cores. The calcified cartilage (C) does not become enveloped by bone. In the metaphysis (M), osteoblasts (arrows) are present but bone collagen deposition is absent. GP = Growth plate.

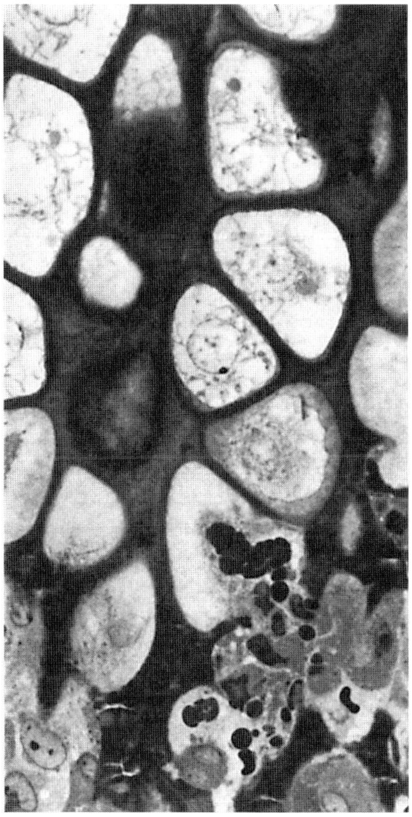

11

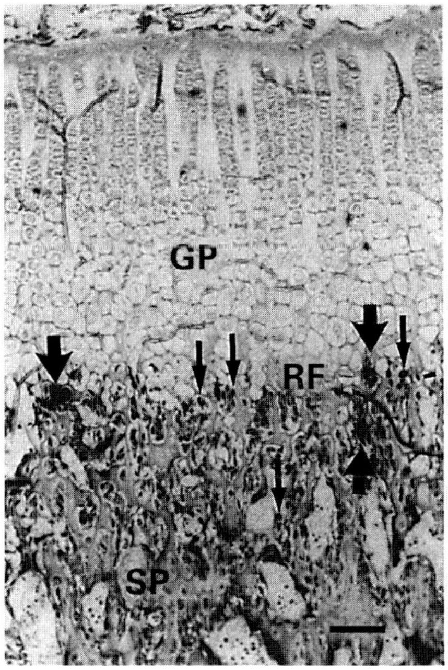

12a

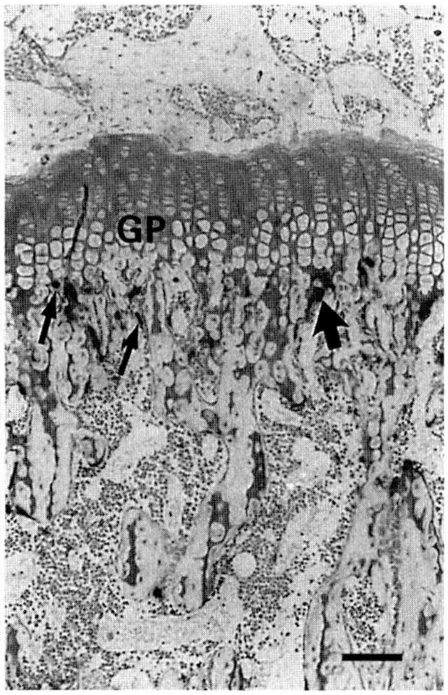

12b

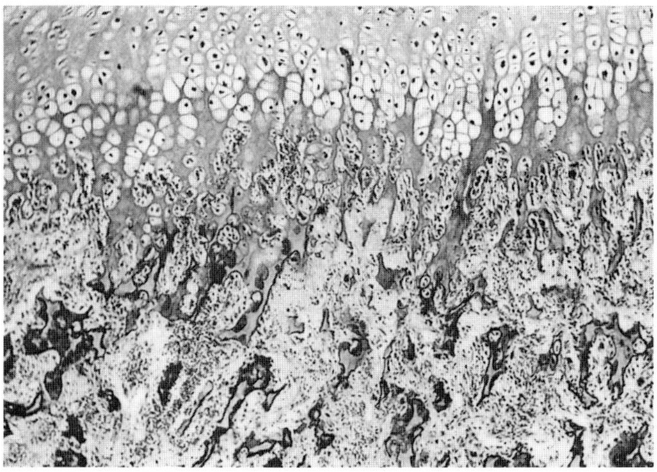

13

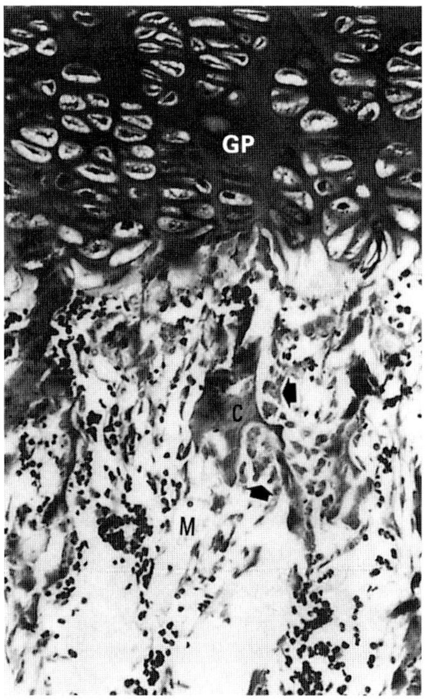

Fusion of Primary and Secondary Ossification Centers

The fusion of an epiphysis with a metaphysis occurs in most long bones during the last year of puberty and implies cessation of the bone's growth. Fusion occurs at different ages for different bones. Fusion starts with fine bridges between the now mature primary and secondary ossification centers **(figure 14)**. Histologically, it is characterized by extension of osteogenesis towards the reserve zone and the gradual elimination of initially the proliferative and later the hypertrophic zone. Matrix calcification proceeds from metaphysis to epiphysis to form osseous bridges that eventually would replace cartilage cell columns. Fusion progresses from the marginal regions centrally, and the cartilage tube would gradually shrink in size until it disappears. Growth of a given bone culminates with the initial bridging, and it may take a whole year before the entire cartilage tube ossifies and calcifies.

Fig. 14
Fusion of primary and secondary ossification centers.
a Photomicrograph demonstrating the proximal end of the radius from a 17-year-old girl. On the left, the plate is still intact, though under higher magnification (**b**) it is clear that enchondral ossification has already ceased. On the right, the degenerating plate has been breached in several places and at the far right end synostosis is evident. × 5.
b Photomicrograph illustrating the remaining inactive, but not yet disrupted growth plate shown in **a**. × 25.
c Skeletal maturation X-ray of a 16-year-old male showing fine bridges between the now mature primary and secondary ossification centers.

Direction of Growth

As a reserve cell is differentiated to become first a chondroblast, then a chondrocyte and then a hypertrophic cell, it changes its orientation relative to its surroundings **(figure 15)** [18]. One can imagine the index cell, as it remains in an unchanging position, while the growth plate grows. Its relative migration from the epiphysis to the metaphysis implies that the former grows away from the latter through regain of proximal elements which replace the index cell. In that respect, growth occurs from the metaphysis to the epiphysis.

Fig. 15
An index cell is pictured in the four zones of the growth plate. As longitudinal growth proceeds, the index cell remains in a constant position, while the growth plate undergoes its developmental stages.

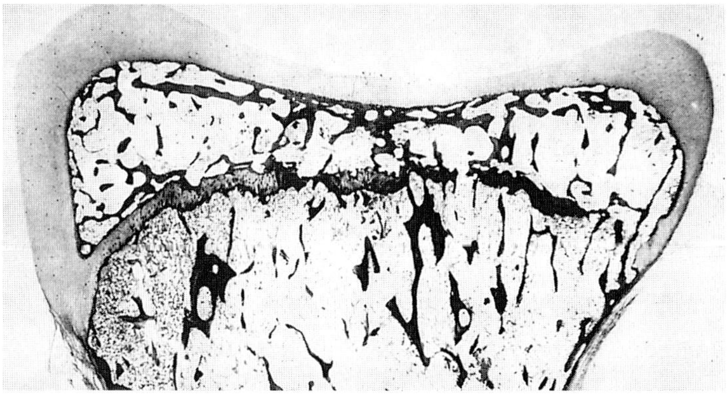

14a

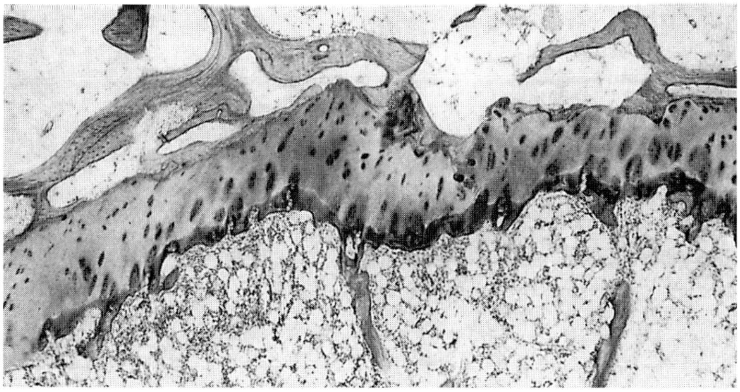

14b

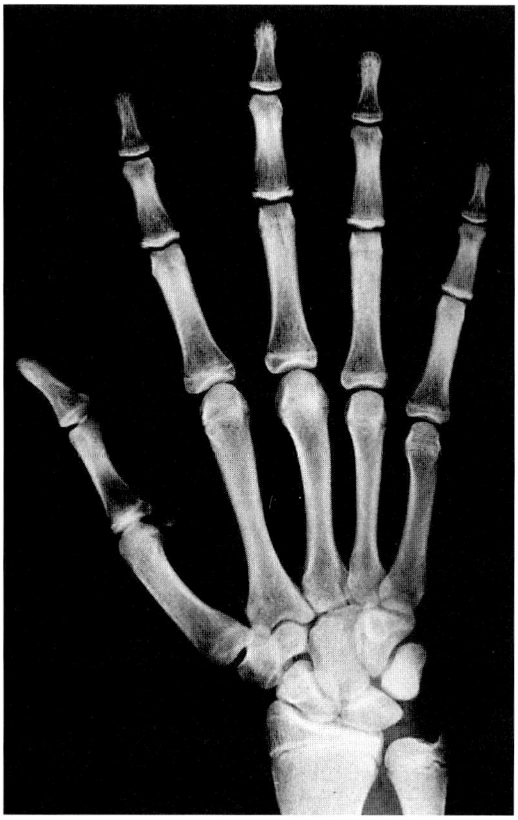

14c

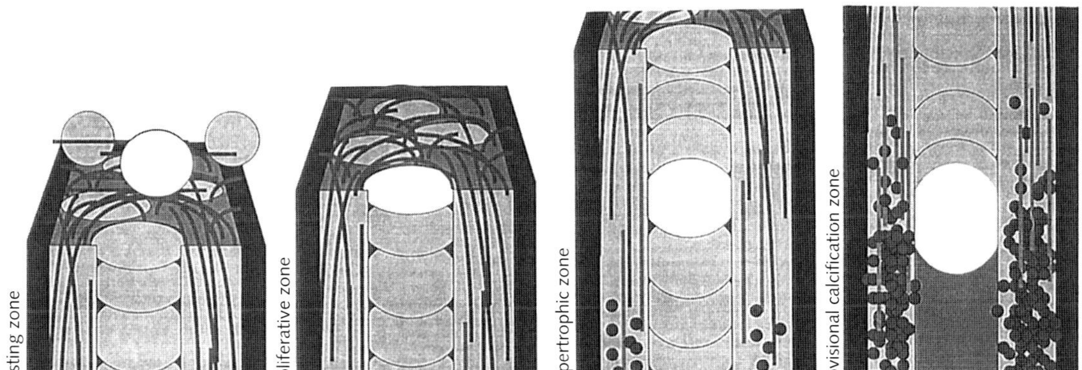

15

Marginal Regions of the Growth Plate

The growth plate has a discoid form, surrounded by a perichondrial ring. Bordering the central part of the plate, limited by the bony epiphysis and metaphysis and adjacent to the perichondrium, the marginal region is made up of cartilaginous tissue, with no secondary ossification center at this level (**figure 16**). Its reserve zone is contiguous to that of the central region. It is wider and has a greater number of cells in the proliferative zone, which are smaller in size compared to the central region [19]. The hypertrophic cells are similar in size and number and undergo similar osteogenesis to that of the central part. In that respect, marginal regions represent pure chondroplasia. The function of this zone is not well specified but it is interesting that its expression of IGF-1 and IGF-2 is higher then that observed at the central region [20], which corresponds to the faster growth in the marginal regions. Chondroblasts of the marginal region can differentiate into hypertrophic chondrocytes, and then undergo apoptosis, or may differentiate into osteoblasts and contribute to subperiosteal ossification.

Fig. 16
Growth of the margins. Photomicrograph of the proximal tibial epiphysis of a 15-day-old rat. A = Central region; B = lateral region. The latter is higher and has a greater number of cells in the proliferative zone that are smaller in size. The borderline cells will differentiate into either hypertrophic chondrocyte or subperiosteal osteoblasts.

Cuboid Bones (Carpals and Vertebrae)

Whereas primary ossification centers of long bones are well ossified at birth, the carpal bones of the hands portray a unique opportunity to observe primary ossification centers in their making. Conceptually, understanding and evaluating the maturation of carpal bones reflects on the maturation of the vertebrae.

A single film will often show the entire range of the maturational phases of the primary ossification center (**figure 17**).

Fig. 17
Primary ossification centers. This X-ray of a 3-year-old boy shows the well-formed ossified and calcified primary ossification centers of the long bones, along with those of cuboid bones, in various states of maturation. The epiphysis of the radius opposes the metaphysis and a growth plate has evolved, reflecting chondrogenesis and osteogenesis. The epiphysis of the 1st metacarpal contains its secondary ossification center and the growth plate has not formed yet, reflecting osteogenesis alone. Some carpal bones have already ossified and calcified and approach their ultimate size, whereas others have just emerged from their hyaline cartilage background.

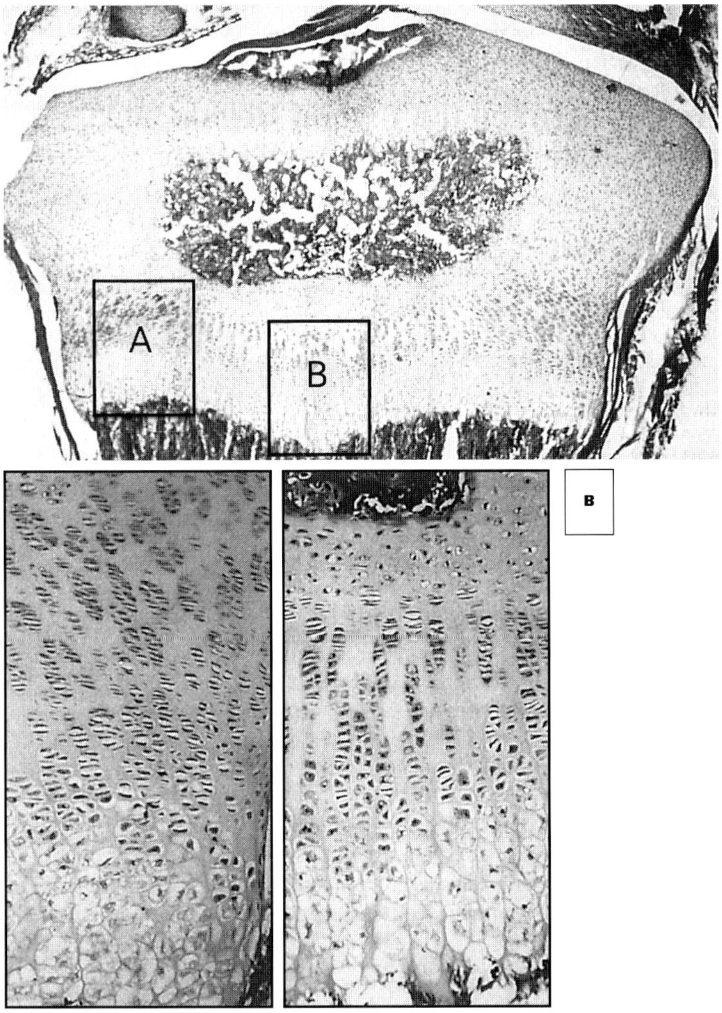

A

B

16

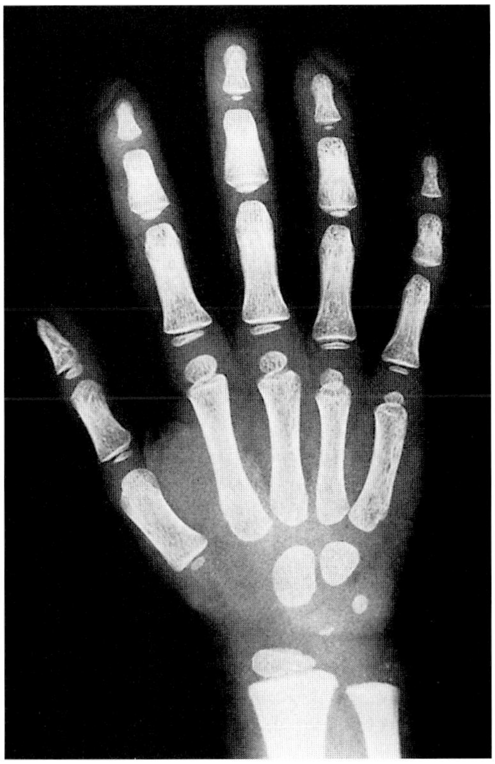

17

Cuboid bones develop through a pure process of osteogenesis **(figure 18)**. Centripetal ossification of the hyaline cartilage occurs through the osteogenetic phases of cartilage hypertrophy and calcium deposition, followed by resorption and formation of bone. Osteoclasts, osteoblasts and growth factors are initially supplied by nutrient vessels, then by the later developed bone marrow. Ossification of another cuboid bone, the vertebra, is seen in **figure 19**, which photomicrographically illustrates the vascularization of the central marrow, with centripetal hypertrophy of chondrocytes, replacing the peripheral hyaline cartilage.

Fig. 18
Osteogenesis of a cuboid bone.
a Photomicrograph of a 5-week-old rat's carpal bone, demonstrating concentric growth of a primary ossification center. × 9.
b Details of **a** showing ossified tissue, nutrient vessel and articular cartilage. ×35.

Fig. 19
Ossification of a cuboid bone, the vertebra. Vascularization of the central marrow, with centripetal hypertrophy of chondrocytes replacing the surrounding hyaline cartilage model. Note the cartilaginous end of the rib on the left. × 10.

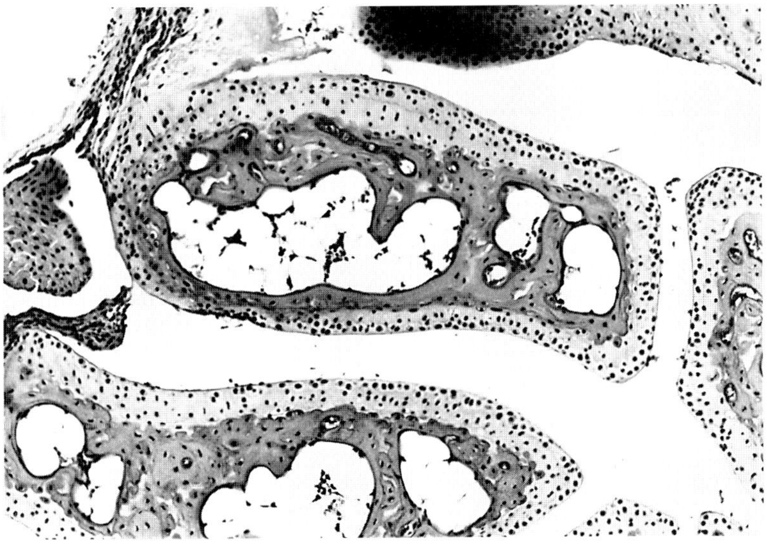

18a

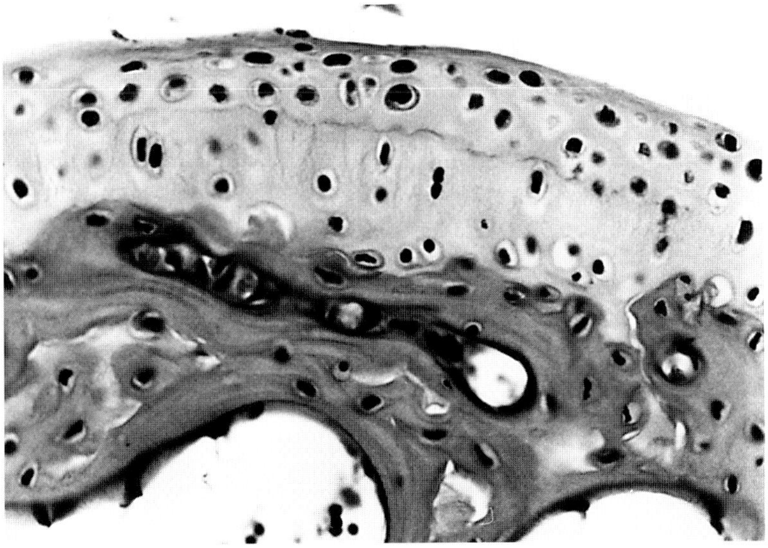

18b

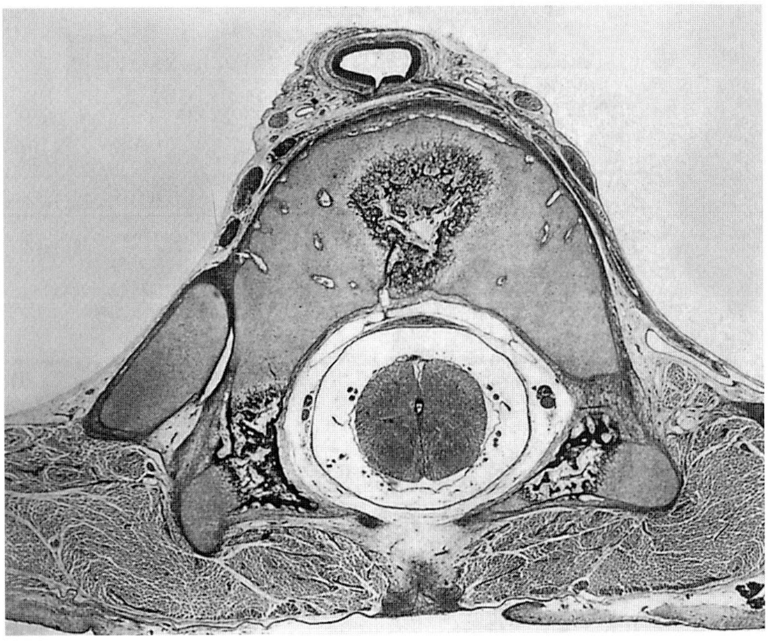

19

Implications for Skeletal Maturity Readings

In all the methods that are utilized for clinical appraisal of skeletal maturity, the user is expected to read each bone separately. This would enable a meaningful evaluation of the underlying mechanisms. Yet, as soon as individual bones are read, the user is then expected to average the scores in some of the methods or sum them up in others. As the TW2 method for evaluation of bone maturity was developed, Tanner et al. [21] realized that the 'long bone' and 'short bone' readings (RUS, for radius ulna and short bones) are often in disparity with the reading of the carpal bones. They indicated that the mean difference between carpals and RUS indices in normal girls is 0 over the range of maturity index 2 – 11, with an SD of 0.7–1.0. Even within the RUS bones, maturity of the short bones of the metacarpals and phalanges often differ from maturity of the radius and ulna. The reasons for these differences lie within the two processes of chondroplasia and osteogenesis that reveal themselves in the hand and wrist film. The appearance in an X-ray film of primary ossification centers of the cuboid bones and of secondary ossification centers in the epiphyses of long and short bones during early childhood reflects the maturational process of osteogenesis. The appearance of the secondary ossification centers of long and short bones during later childhood and adolescence reflects a combination of osteogenesis and chondroplasia [6]. For an insight into the mechanisms behind the film, minimal effort is required to group the individually read bones to RU – radius and ulna –, S – short bones –, and C – cuboid carpal bones. This simple tool allows some insight into a possible aberrance of the basic processes of osteogenesis and chondroplasia. The maturity scores of carpal bones often resemble the RU score and the RUC score is more homogeneous than the RUS score. Understanding the control of these two processes allows inferences on the underlying pathology to be made from the X-ray film.

Mostly, the clinician is not interested in the maturation of bones of the hand and wrist. Bone maturity films are expected to reflect on those distant bones that contribute to the child's growth. It is commonly felt that maturity of the cuboid carpal bones may stand for maturity of the cuboid vertebrae and that maturity of the short bones of the hand may stand for maturity of the femur and tibia. Yet, these logical assumptions still require scientific confirmation.
The advantage of the hand-wrist area is that a single film covers 30 different ossification centers, which represent the various growth patterns.

Radiographic Perception

Radiographic Perception

Unless special action is taken, the X-ray film will mostly only show the calcified tissues. In the normal hand and wrist, these include not only the bone but also the calcified cartilage. **Figure 20** depicts soft radiograms of the distal femur of an infant and a young boy and their parallel sagittal cuts. The secondary ossification center at the infantile epiphysis undergoes an identical process of osteogenesis to that of a primary ossification center. A growth plate has not yet formed across the metaphysis. Calcified matrix is seen in the radiogram as a distinct wide radiopaque layer.

Fig. 20
The radiographic perception. Distal femur at ages 3 months (**a, b**) and 3 years (**c, d**) in sagittal cuts (**a, c**) and their parallel X-rays (**b, d**). The secondary ossification center of the infantile epiphysis undergoes an identical osteogenesis process to that of a primary ossification center and a growth plate has not yet formed across the metaphysis. The arrows in **a** indicate a faint zone of provisional calcification, shown as a delicate radiopaque line between the epiphysis and the shaft. At age 3 years, as the secondary ossification center widens, still surrounded by a boundary of proliferating cartilage, a growth plate forms and stands out clearly between the epiphysis and the metaphysis. Calcified matrix is seen in the radiogram as a distinct wide radiopaque layer.

Demarcation Line

Whereas hyaline cartilage of the bone anlage and the proliferating zone are translucent to X-rays, calcified matrix and bones are visible. Thus, the boundaries of a tissue that would be visible on a radiogram of a growth plate are calcified bone trabeculae at the epiphyseal end and the calcified matrix of the hypertrophic chondrocytes at the metaphyseal end. In **figure 20**, calcified matrix is seen in the radiogram as a distinct wide radiopaque layer. **Figure 21** is a scanning electron microscopic (EM) depiction of the vaporous calcified matrix delineating the demarcation of an eventual radiological film through the emptied lacunae. The organic material has been digested and gold plating allows for EM demonstration of the mineral skeleton. It is clear that the demarcation line of a radiogram delineates the border between calcified and uncalcified cartilage.

Fig. 21
The demarcation line. Growth plate of the rat tibial epiphysis. This inorganic specimen shows the vaporous appearance of the mineralized matrix, which surrounds the hypertrophic cells, delineating the demarcation of an eventual radiogram through the emptied lacunae.

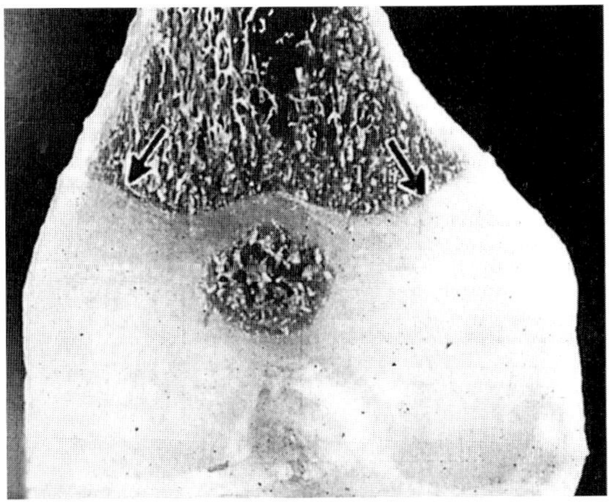

20a

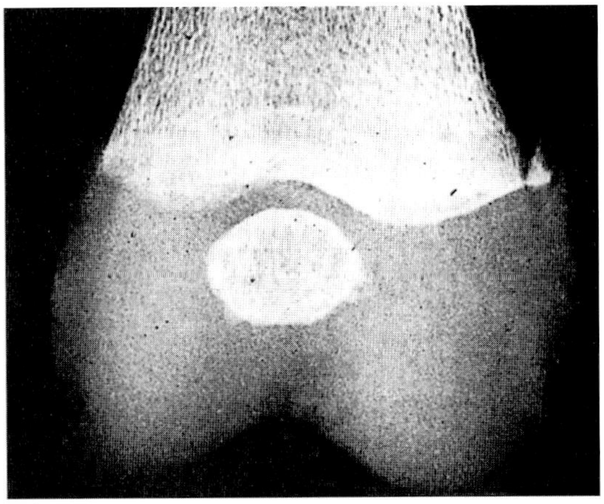

20b

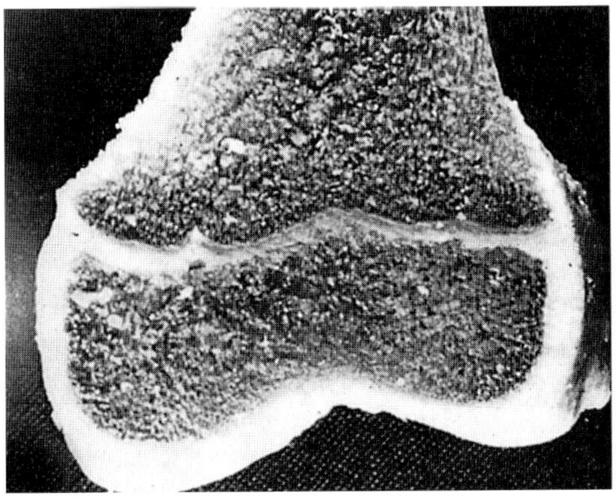

20c

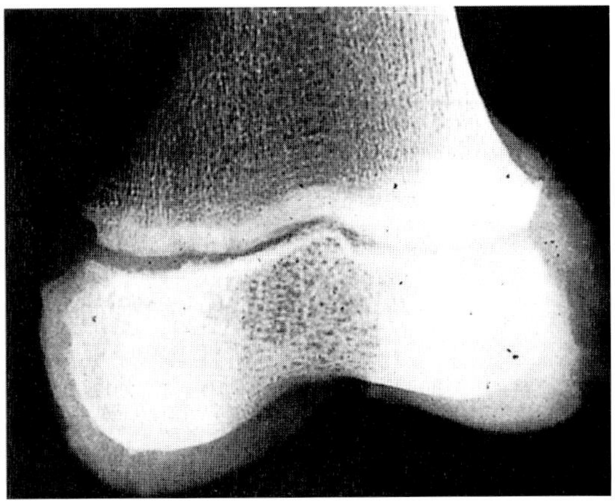

20d

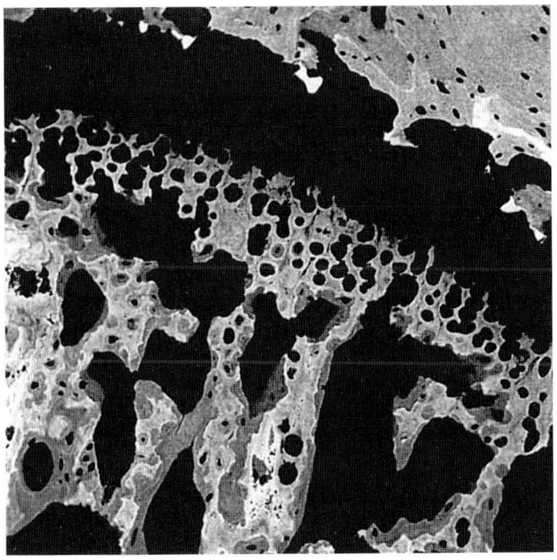

21

Such a view through the growth plate provides a unique opportunity to assess conditions of disturbed calcification. **Figure 22** shows a typical X-ray of a child with rickets. The thick radiopaque layer of calcified cartilage is obvious. It reflects on increased matrix calcification by rachitic hypertrophic chondrocytes. The ends of the metaphyses are splayed because of an increase in the appositional growth at the marginal regions. These changes occur in response to retarded or absent enchondral bone growth.

Thus, the demarcation line of a radiogram separates chondroplasia, with its uncalcified cartilage, from osteogenesis, with its calcified cartilage and resorption front.

Fig. 22
Rickets. This 6-year-old boy suffers from vitamin D-dependent rickets type 2, due to defective vitamin D receptor. The thick radiopaque layer of calcified cartilage is best seen in the distal ulna, which provides a glimpse into hyper-calcification of matrix by rachitic hypertrophic chondrocytes. The ends of the long bones are splayed because of an increase in appositional growth in response to retarded or absent enchondral bone growth.

Chondroplasia and Osteogenesis

In the consecutive phases of bone maturation, the radiographic elements represent the underlying mechanisms. Thus, primary ossification centers of the long bones and cuboid bones represent sheer osteogenesis. A bone and wrist radiogram of a young infant, which shows no secondary ossification centers, constitutes a net view of osteogenesis **(figure 23)**.

Fig. 23
X-ray views of osteogenesis and chondroplasia. Radiograms of a healthy girl at 14 months of age **(a)** and later at 21 months of age **(b)**. As an infant her hand and wrist X-ray film shows primary ossification centers of two carpal bones and secondary ossification centers of metacarpals and proximal phalanges. This provides a view of unqualified osteogenesis. Several months later, as the secondary ossification center of the long and some of the short bones are seen, and as some other secondary ossification centers grow and approach the primary ossification center, one has an opportunity to glimpse at both osteogenesis and chondroplasia.

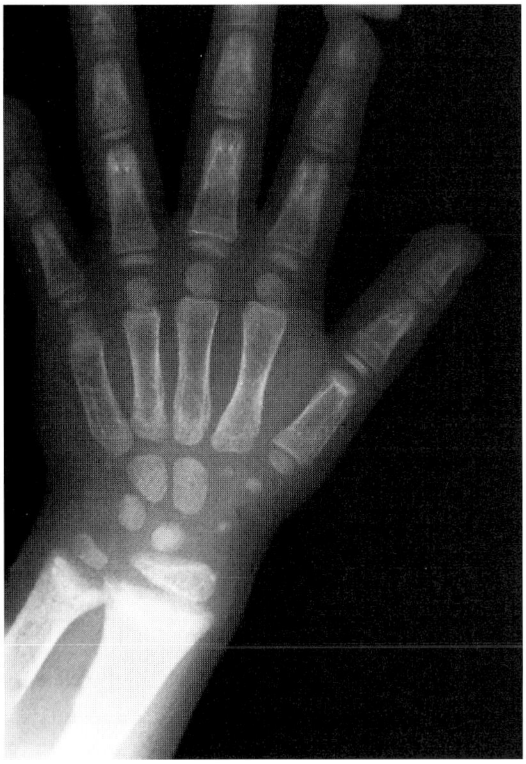

22

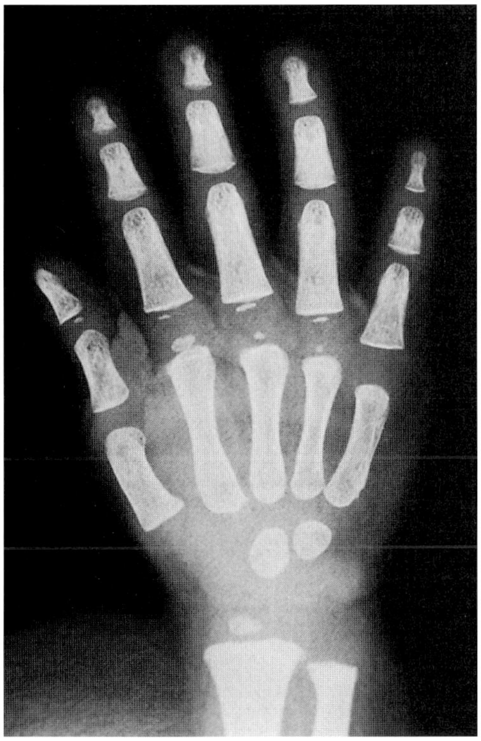

23a

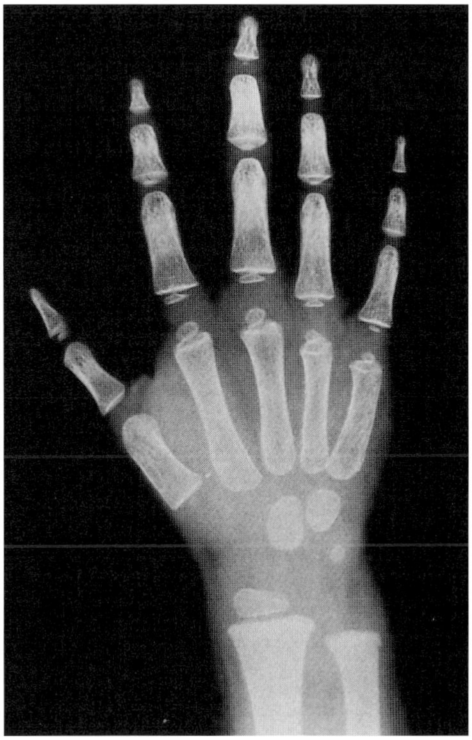

23b

As growth plates have not formed yet, such X-ray films portray osteogenesis only and would be delayed in pathophysiological processes of impaired osteogenesis. Only after the age of 3 years are some epiphyses large enough to enable depiction of views of the growth plate, and able to allow the viewer a glimpse at the chondroplasia **(figure 24)**. Over the next stages of maturity, the radiograms show parallel views of the epiphyses as they verge on the metaphyses; a growth plate forms and provides a glimpse at the chondroplasia.

Clinicians recognize defects in these two processes, which contain the term chondroplasia and osteogenesis. Achondroplasia is a severe defect of bone formation **(figure 24)**, yet the pathophysiology of this disease reaches beyond malfunction of the growth plate. As a result of a defective receptor for the fibroblast growth factor (FGF), chondrocyte columns are shorter and lose their parallel orientation, resulting in short and wide bones. Osteogenesis of the carpal bones proceeds unaffected. In the absence of enchondral bone growth, the metaphyses spread out at the marginal regions. The ulna is shorter than the radius. The hands, which do not bear any weight, are less affected than the legs. Yet, achondroplasia is not only a disease of the growth plate. In fact, its impact is evident even during the formation of primary ossification centers and long before the appearance of any epiphyseal growth plate. A photomicrograph of a typical femur is shown in **figure 25**. Taken from a stillborn premature fetus, the bony shaft is abnormally short relative to the size of the cartilaginous epiphyses, indicating that even before a growth plate has formed, bone maturation and growth are severely disturbed. In another specimen from a 7-month-old achondroplastic female, the distal end of the femur shows the bone trabeculae directly opposing the cartilage epiphyses with no evidence for enchondral ossification.

Fig. 24
Achondroplasia. As a result of defective receptor for the fibroblast growth factor (FGF), chondrocyte columns are shorter and lose their parallel orientation, resulting in stunted and wider long and short bones of the hand and wrist. Osteogenesis of the carpal bones proceeds unaffected. In the absence of enchondral bone growth, the metaphyses spread out at the marginal regions. Primary ossification centers of the carpals and secondary ossification centers are poorly ossified and look transparent with paper-thin cortex. The ulna is shorter than the radius.

Fig. 25
Achondroplasia.
a Photomicrograph of a femur from a stillborn premature fetus The bony shaft is abnormally short relative to the size of the cartilaginous epiphyses.
b Photomicrograph of another 7-month-old achondroplastic female, the distal end of the femur shows the bone trabeculae opposing directly the cartilage epiphyses with no evidence for enchondral ossification. × 9.
c Photomicrograph of the distal femur from the same patient shown in **b**, illustrating the epiphyseal metaphyseal interface. There is no evidence for active enchondral ossification. Trabeculae of bone are opposed directly against the cartilage. × 4.5.
d Photomicrograph of the infant shown in **b** and **c**, illustrating the costal growth plate. The straight columns architecture is lost and matrix has not calcified between the hypertrophic cells. × 58.

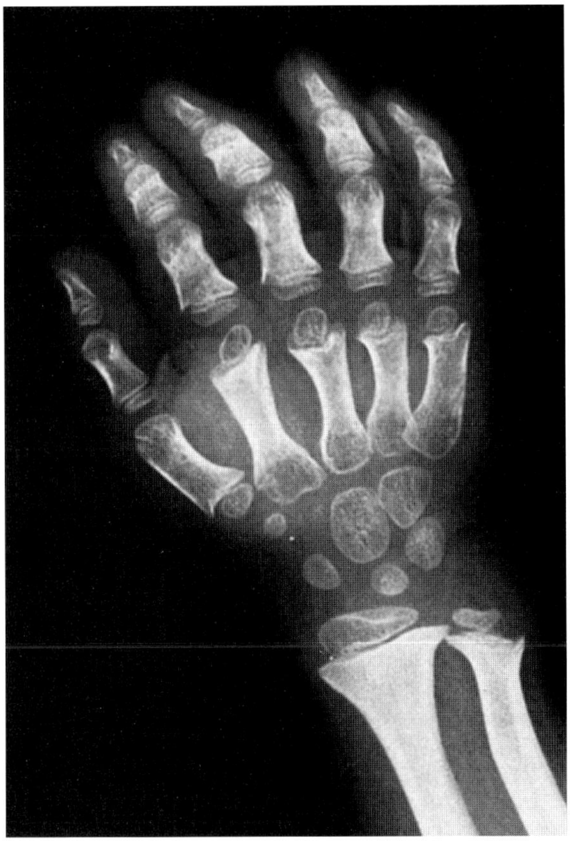

24

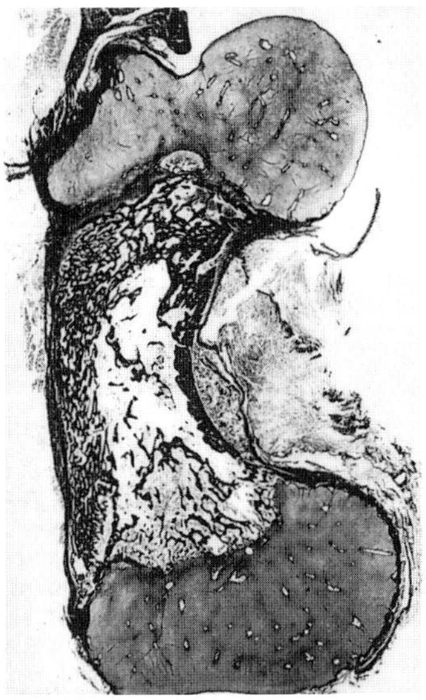

25a

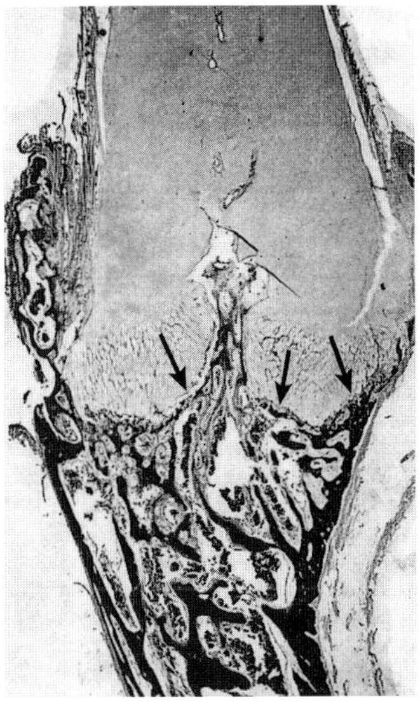

25b

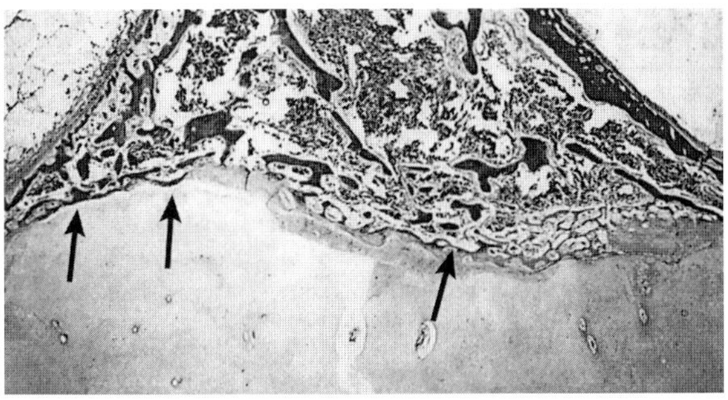

25c

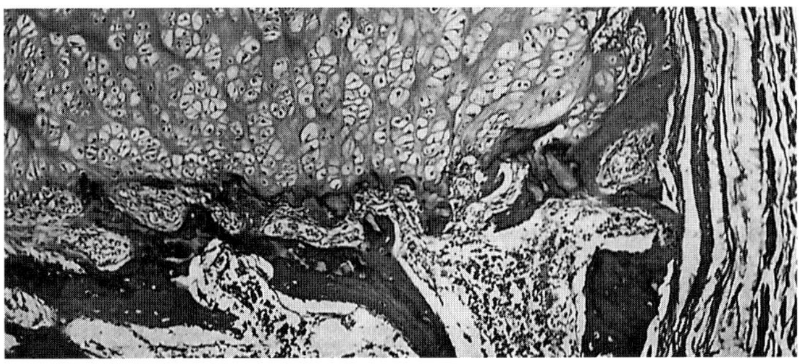

25d

In the milder form of hypochondroplasia, long and short bones are proportionately short as compared to the trunk, but the configuration is normal and osteogenesis proceeds unaffected **(figure 26)**. This is a difficult diagnosis as it relies mostly on disproportionately stunted long bones. As the long bones of infants and toddlers are short anyway, diagnosis is often delayed.

Impaired osteogenesis is seen in osteogenesis imperfecta (**figure 13**), hypothyroidism and in osteopetrosis **(figure 27)**. In the case of failure of resorption of the osteopetrotic calcified cartilage, the tissue does not transform into bone. The column organization of the hypertrophic zone is disturbed. Prevention of modeling results in flared metaphyses and wide shafts. The metacarpals are described as a 'bone within a bone' [22].

At the other end of this spectrum, enhanced osteogenesis and rapid bone turnover are evident in hyperphosphatasia **(figure 28)**. This autosomal-recessive disease of deossification is referred to as Paget's disease of the young. The clinical picture is dominated by bowing of the long bones. The cortices are dense but thin and are associated with loose trabeculation of the long bones, which forms a parallel layer of bone resulting in loss of the concentric progression of osteogenesis at the metacarpal epiphyses and thus their square-shaped appearance. The resemblance of this configuration to that seen in Jansen-type metaphyseal chondrodysplasia suggests a role for PTH/PTHrP in this disorder.

Fig. 26
Hypochondroplasia in a 9-month-old infant.
a The limbs are short as compared to the normal-sized trunk. The long bones are proportionately short but their configuration is normal.
b The short bones are only mildly shorter than normal and osteogenesis proceeds unaffected.

Fig. 27
Osteopetrosis. With failed resorption of the calcified cartilage, the tissue does not transform into bone. The column organization of the hypertrophic zone is disturbed. Prevention of modeling results in splayed metaphyses and wide shafts. The metacarpals are described as a 'bone within a bone'.

Fig. 28
Hyperphosphatasia. This autosomal-recessive disease is referred to as Paget's disease of the young. Cortical thinning is associated with parallel loose trabeculation of the bone, which results in loss of the concentric progression of osteogenesis at the metacarpal epiphyses and in their square appearance.

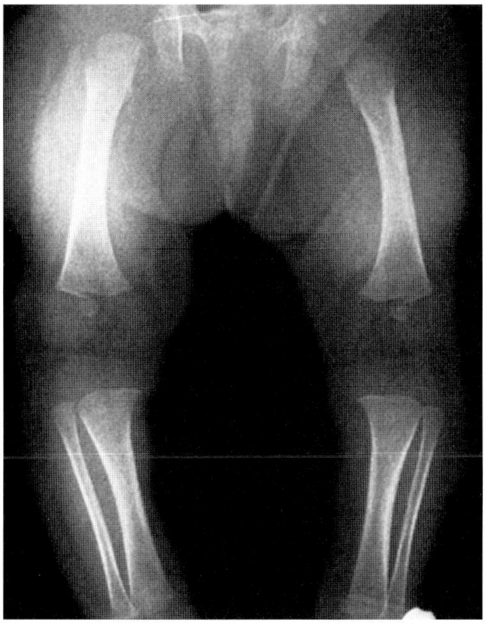

26a

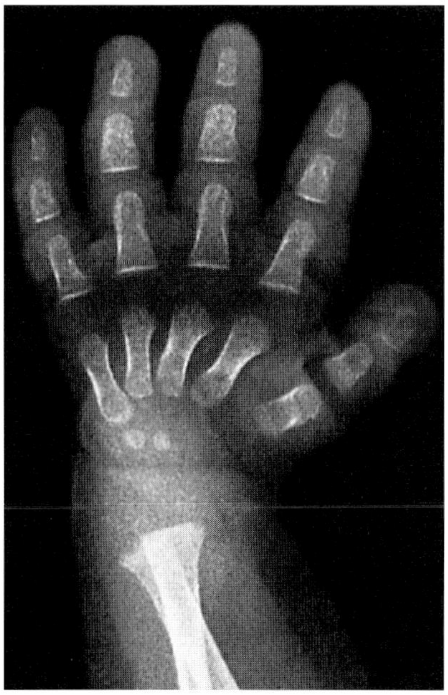

26b

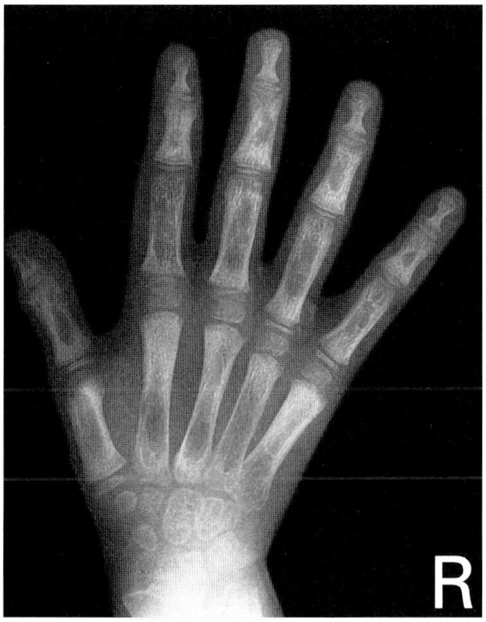

27

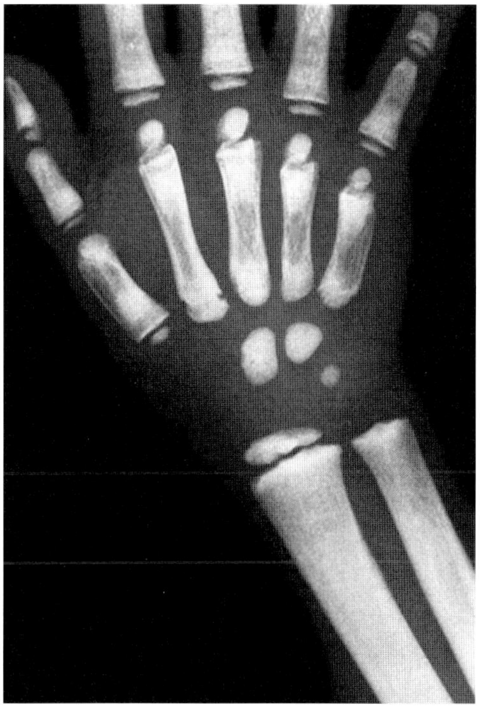

28

Growth Arrest Lines

The history of a child's growth is often evident from changes in the rate of longitudinal bone growth, visible as growth arrest lines, also known as Harris growth lines. These may be found in several bones or in a single bone. Transverse lines of increased radiodensity are confined to the metaphyses and designate transverse bone trabeculae, perpendicular to the regular longitudinal orientation of the trabeculae. The distance of a growth arrest line from the growth plate indicates the time elapsed since growth arrest and several such lines give a history of the long-term growth profile. **Figure 29** shows an X-ray of a distal femur removed from a 2.5-year-old child who died after surgery for hydrocephaly. The two transverse lines are seen in the photographic section (indicated by arrows) and in the photomicrograph, showing two rows of trabeculae that are disposed in an irregularly transverse pattern, resulting from periods of growth arrest. A more subtle case is presented in **figure 30**. This 13-year-old boy with short stature suffered a psychosocial insult 2 years previously. His mother left the family for an unknown destination and resurfaced 6 months later. The pediatrician documented a growth rate of 1 cm/year over that period and a catch-up growth thereafter. Thus, a relatively short systemic disease may give rise to a single line seen throughout the skeleton, whereas a growth arrest line in a single bone usually implies a local insult, such as a fracture. Over long periods of time, the growth arrest line remodels and fades.

Fig. 29
Growth arrest lines.
a Radiograph of a distal femur removed from a 2½-year-old child who died after surgery for hydrocephaly.
b The two transverse lines are seen and are indicated by arrows.
c Photomicrograph showing two rows of trabeculae that are disposed in an irregularly transverse pattern, resulting from periods of growth arrest. × 2.

Fig. 30
Psychosocial deprivation. This 13-year-old boy with short stature suffered a psychosocial insult 2 years previously. The middle phalanges show a single sclerotic line that represents an insult that continued for 18 months.

29a

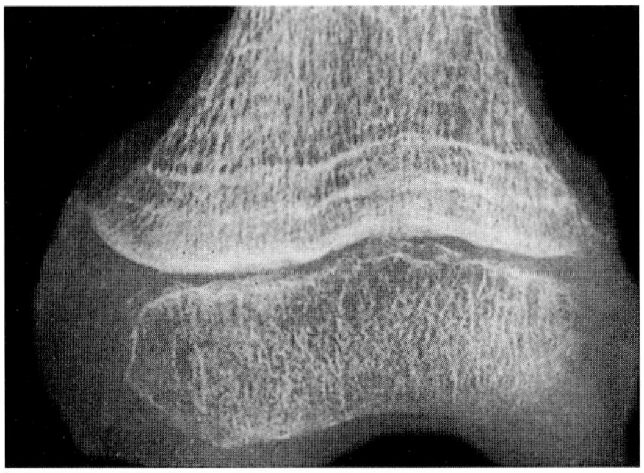

29b

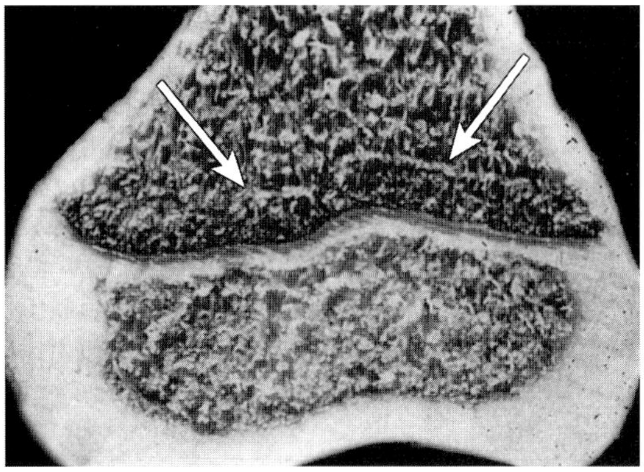

29c

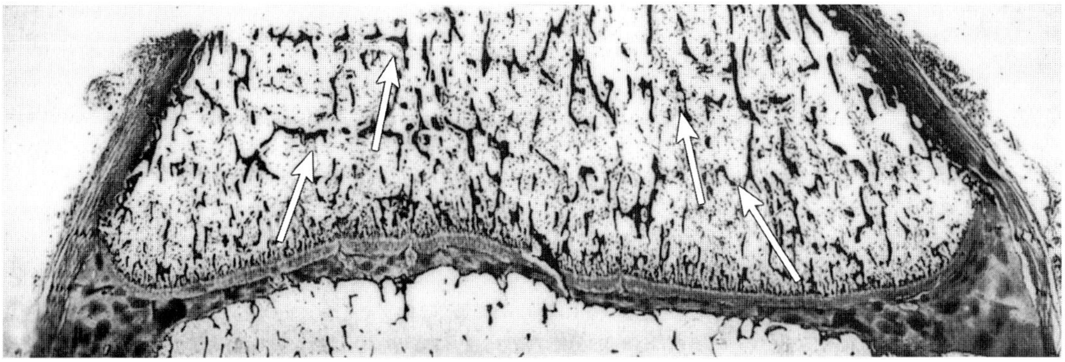

30

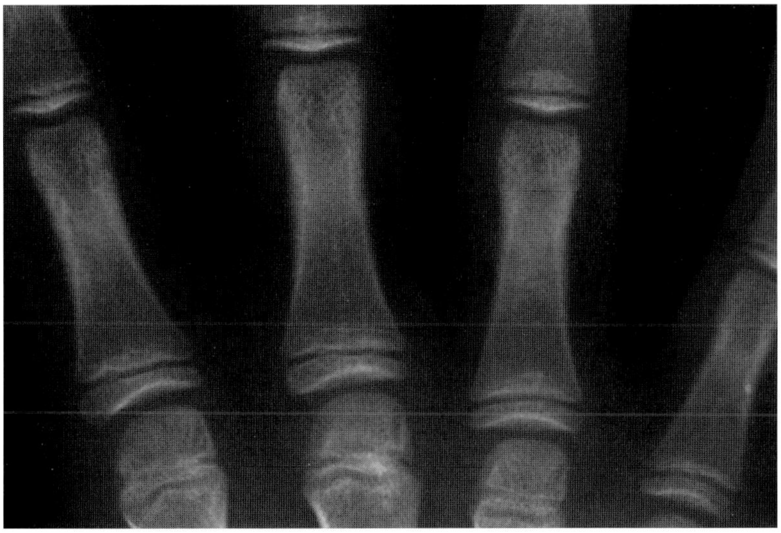

Normal Variations

The density of phalangeal epiphyses may vary among bones in the same film. Hyperdensity of a single epiphysis indicates local sclerosis, mostly with no pathological significance, or may result from current problem or a history of local arthritis or chondritis.

Some cuboid bones may show two ossification centers that would ultimately fuse. Adjacent carpal bones may fuse into a single atypical larger bone. Carpal bones may show unusual disfiguring bony processes or irregular mineralization. All these are more common in girls than boys [23].

False ossification centers may appear in the proximal ends of the metacarpals, where, normally, no ossification centers are present **(figure 31)**. These are spikes of bones, separated from the proximal shaft by a notch. They do not represent epiphyses, nor do they indicate growth plates. They are to be distinguished from accessory secondary ossification centers, which undergo similar histological and radiological growth to regular epiphyses, and fuse later with the adjacent metaphyses **(figure 6)**.

Missing epiphyses are sometimes seen, mostly in the middle phalanges, and are more frequent in girls than in boys.

Fig. 31
False ossification centers. Hand radiogram of a 3 years and 5 months old boy who presented with short stature. The long bones index is $1^{2}/_{12}$, the cuboid bones index is $^{10}/_{12}$ and the short bones index is $2\ ^{5}/_{12}$. This combination suggests a delay in osteogenesis. These spikes of bones (arrow), separated from the proximal shaft by a notch, do not represent epiphyses, nor do they develop growth plates.

Maturational Sequence

Bones mature in an orderly fashion. The sequence of bone ossification and calcification allows for unique assignment of an index to the radiographic appearance of each bone. A deviation from these orderly events can be used clinically to identify an insult to bone maturity retrospectively. A severe systemic disease, malnutrition or an endocrine malfunction may prevent the ossification of a primary or secondary ossification center. Thus, the age at onset of malnutrition or maternal deprivation that should have appeared at that age can be read from missing ossification centers years after recovery.

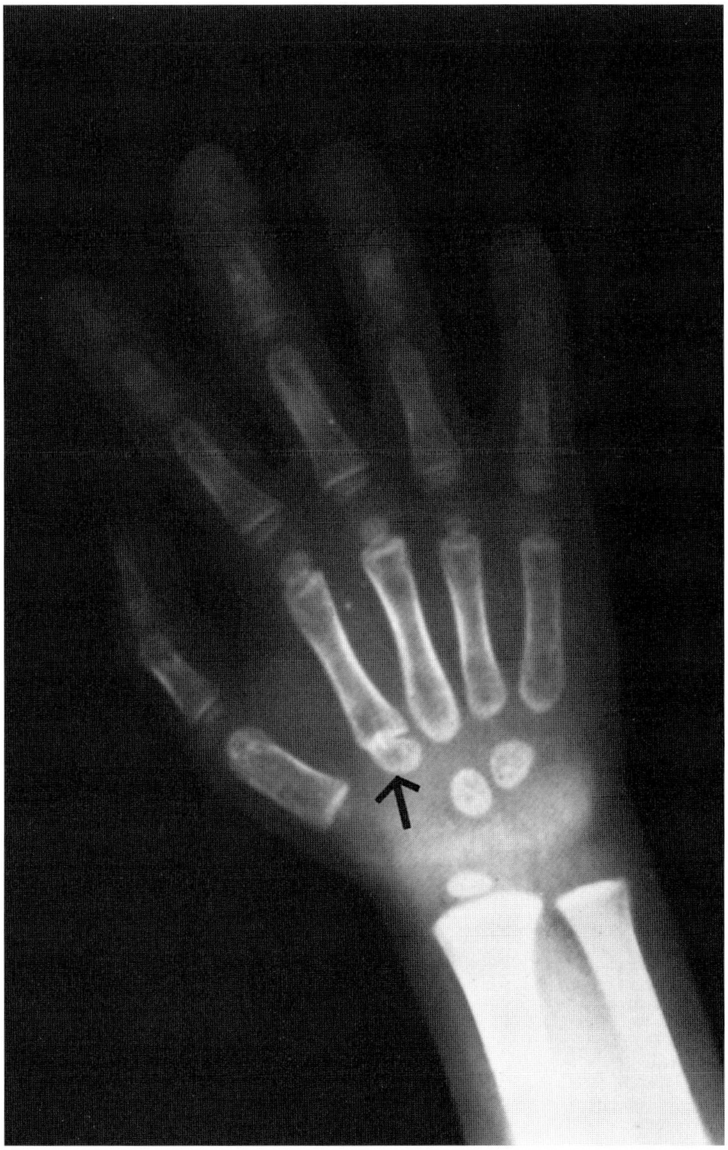

31

Other Imaging Tools

Ultrasonography provides a new imaging tool to observe the structural relationship of epiphyses, metaphyses and growth plates. Victor Bialik from Haifa, Israel, has developed a new measure of newborn bone maturation that is based on ultrasonography of the distal femur epiphysis. **Figure 32** shows ultrasonographic imaging of the distal femur growth plates in an infant and in a prepubertal child. The relationship of epiphysis and metaphysis is well represented. This promising technique requires further development.

Magnetic resonance imaging (MRI) shows a striking appearance of the epiphyseal growth plate and the secondary ossification centers. MR imaging can successfully evaluate the cartilaginous structures of the developing skeleton. This is best demonstrated in an extreme case of x-linked hypophosphatemic rickets **(figure 33)** [24].

Fig. 32
Ultrasonography of the distal femur in an 8-month-old female (left) and a 6-year-old male (right). The dashed line denotes the supposed complete figuring of the bone.

Fig. 33
MR imaging of rickets. Metaphyseal widening and high signal of calcified cartilaginous matrix (arrows) in the distal tibia of a 13-month-old girl with nutritional rickets.

32

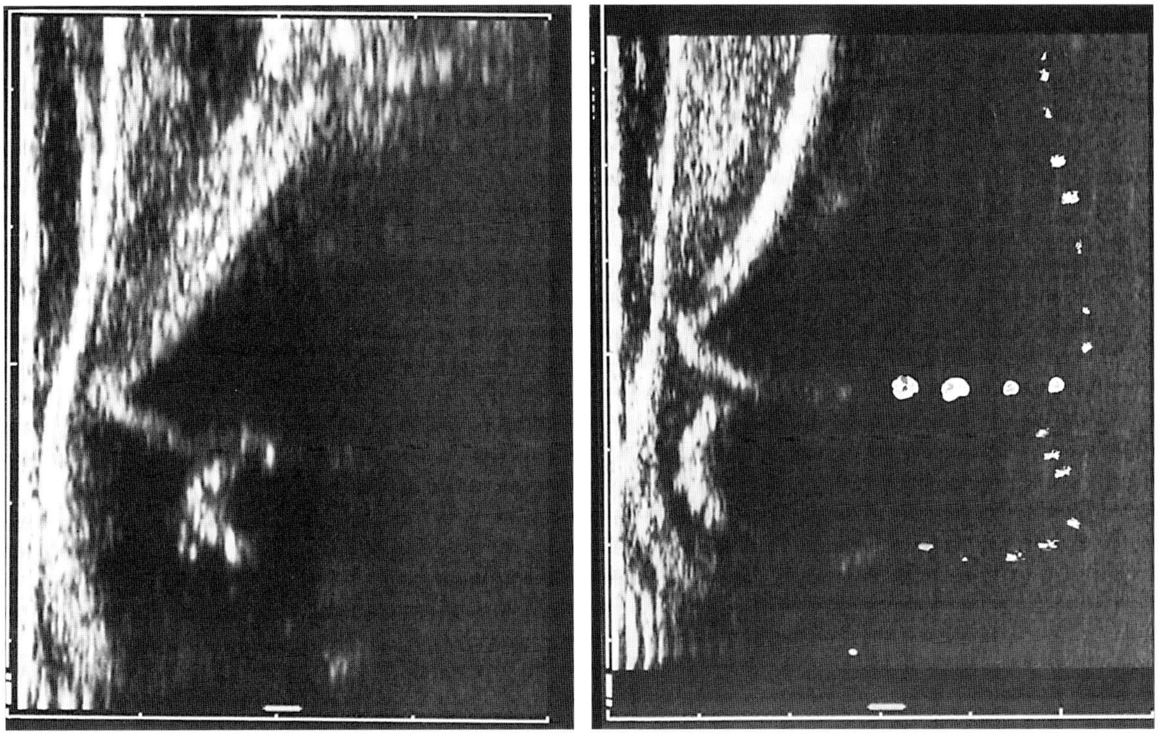

33

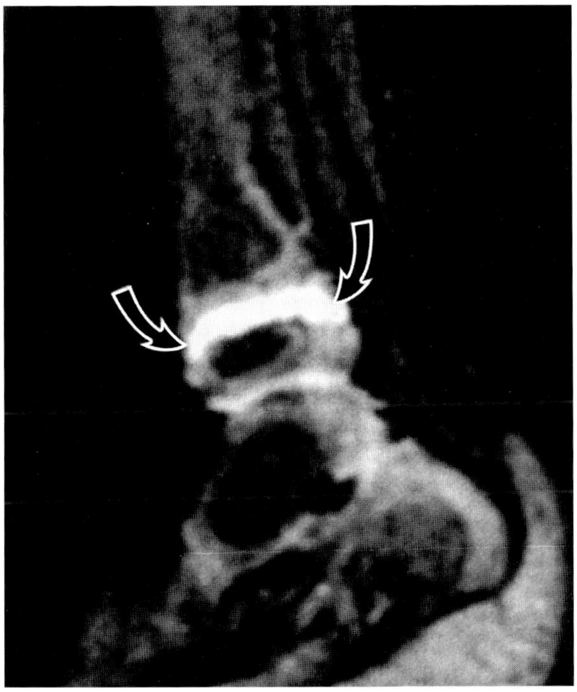

Endocrine Regulation of Bone Maturation

Growth Hormone – IGF-1 Axis

Physiological Considerations

In 1923, Dott and Fraser [25] injected pituitary extracts to hypophysectomized animals, demonstrated its stimulatory effect on growth, and suggested for the first time a role for a pituitary GH in longitudinal bone growth. GH-deficient children respond to GH therapy with acceleration of their growth rate, due mostly to longitudinal growth of long bones. Both GH and insulin-like growth factor 1 (IGF-1) have direct effects on the growth plate **(table 1)**. Surrounding the growth plate, GH also influences bone itself, connective tissue, vasculature and the bone marrow, and injections of either GH or IGF-1 to hypophysectomized rats stimulates growth plate chondrocytes at all stages of differentiation. The effect of GH is greater than that of IGF-1, but none completely restore the hypophysectomized animal's growth [6].

Reserve chondroprogenitor cells express GH receptors [26], and respond to GH by differentiation into chondrocytes. The direct effects of GH and IGF-1 are best exemplified by the study of Maor et al. [27]. They isolated reserve cells from a newborn mouse mandibular condyle, and showed that in vitro incubation with either GH or IGF-1 enhanced differentiation of chondroprogenitor cells into cartilage within 3 – 6 days **(figure 34)** [28].

GH and IGF-1 markedly enhance DNA synthesis, expression of cartilage-specific type II collagen, proteoglycan, chondrocalcin and 100-kD protein. The reserve zone developed into a nodule of differentiated chondrocytes, including a well-developed perichondral layer. Thus, both GH and IGF-1 accelerate the maturational process of chondroplasia. Yet, GH also induces osteogenesis. It transforms the cartilaginous nodule into bone. These in vitro experiments were supported by an in vivo study, delivering GH or IGF-1 into the growth plate. GH but not IGF-1 stimulated thymidine uptake to the reserve cells [29]. The transition from reserve cell zone to proliferating cartilage is associated with expression of the highest concentration of IGF-1, IGF-1 receptors and the cell's responsiveness to IGF-1. GH stimulation of proliferation seems, therefore, to be an indirect effect of reserve cells that were stimulated by GH to advance into their proliferative phase. With chondrocyte hypertrophy in focus, Hunziker et al. [7] showed conclusively that in vivo injections of GH and IGF-1 stimulate the longitudinal orientation of hypertrophic cells, thereby adding substantially to bone length. At the same time, both GH and IGF-1 have a role in matrix regulation. In the cartilage-bone interface, GH stimulates mobilization of osteoclasts to the resorption front [12]. Similarly, IGF-1 elicits a significant increase in osteoclastic acid phosphatase activity [30]. In

Table 1. GH and IGF-1
A summary of the effects of GH and IGF-1 on the growth plate and its adjacent tissues

	Growth hormone	**IGF-1**
Reserve cells – progenitor to chondrocytes – differentiate into chondrocytes	– express growth hormone receptors – differentiation into chondrocytes – stimulate thymidine uptake	
Proliferating chondrocytes – longitudinal cell division		– express IGF-1 – express IGF-1 receptors – proliferation response to IGF-1
Hypertrophic chondrocytes – cell hypertrophy – longitudinal orientation – apoptosis	– express growth hormone receptors – longitudinal orientation of cells	– express IGF-1 – longitudinal orientation of cells – slows apoptosis
Cartilage-bone interface – lacuna in the calcified matrix – bone marrow invasion – resorption	– marrow invasion – mobilization of osteoclasts – growth hormone binding and activity in osteoclasts – differentiation of bone	– marrow invasion – mobilization of osteoclasts – IGF immunoreactivity in osteoclasts
Extracellular matrix – cartilage matrix – calcification of matrix – bone matrix		– expression of cartilage matrix elements – ^{35}S incorporation into cartilage proteoglycans
Bone marrow – differentiation of lineages – lineages maturation – cell proliferation	– expression of growth hormone receptor in monocytes/macrophages – proliferation of erythro – and granulopoiesis – maturation of erythro – and granulopoiesis	– expression of IGF-1 in monocytes/macrophages – expression of IGF receptor in monocytes/macrophages – proliferation of erythro – and granulopoiesis – maturation of erythro – and granulopoiesis
Bone – resorption and formation – bone turnover and remodelling – bone mineral density (BMD) – serum markers	– stimulates bone remodelling – shorter resorption and reversal time – BMD low in GHD and increase with growth hormone therapy – serum markers low in GHD and increase with growth hormone therapy	– synthesis in osteoblasts – receptors in osteoclasts – stimulation of bone matrix apposition – support of osteoclast formation and activation

the acid phosphatase-positive osteoclast, IGF-1 immunoreactivity but not IGF-1 mRNA was detected.

Observations on the effects of GH and IGF-1 on bone metabolism are relatively recent and appeared in conjunction with studies on GH therapy to adults with GH deficiencies [31]. Utilizing measurements of bone mineral density, bone markers, such as urinary pyridinoline, osteocalcin and a propeptide of collagen type I as well as histomorphometric indices, GH stimulates bone turnover and remodeling. The enhanced bone turnover results from a shorter resorption and reversal time and increases bone formation. More recent evidence suggests that osteoclasts possess IGF-1 receptors [32], which are involved in osteoclast mobilization by GH and in the effects of IGF-1 on bone resorption. The osteoblast mediates the stimulation by IGF-1 and IGF-2 of osteoclast formation and function [33].

Taken together, in the frame of our classification of enchondral ossification, this experimental work shows that GH and IGF-1 have clear stimulating effects on chondroplasia, but also a major role in osteogenesis, where GH has a more dominant effect. The effects of GH and IGF-1 on these two processes are partly interlined and partly independent.

Fig. 34
GH and IGF-1 effects on chondroplasia and osteogenesis.
a Photomicrograph of chondroprogenitor cells that were detached from the mandibular condyle of a newborn mouse. × 143.
b Photomicrograph of a control explant after 6 days in culture. The tissue attained the form of a condensed nodule; its inner portion reveals signs of cartilage. × 143
c Similar conditions to that shown in **b**, but obtained from a GH-treated specimen. The GH-treated culture is 4 times larger than its control and the newly developed cartilage increased in size 5-fold. The periphery has transformed into a bone with multinucleated giant cells that resemble osteoclasts or chondroclasts and express bone specific antigens (not shown).
d Similar conditions to that shown in **a**, but obtained from an IGF-1-treated specimen. The size is similar to that obtained after GH incubation, but no osteogenesis is apparent.

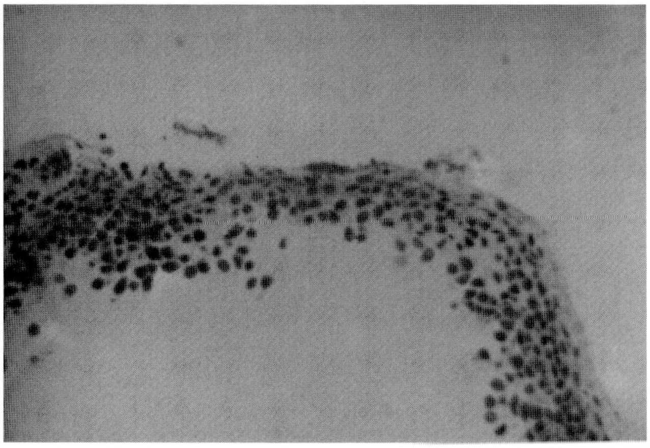

34a

34b

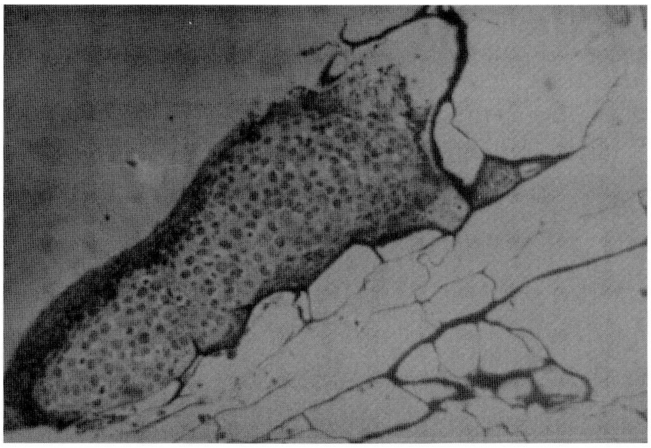

34c

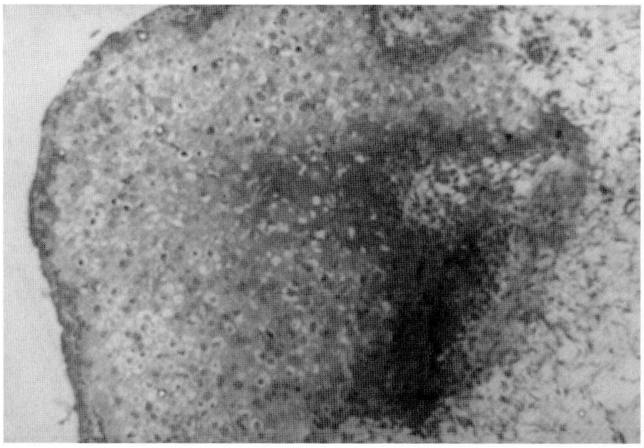

34d

Implications for Skeletal Maturity Readings

As GH is a major stimulator of osteogenesis, and to a smaller extent also of chondroplasia, GH deficiency impairs these processes in that order of severity. **Figure 35** shows the skeletal maturity of a boy with GH deficiency. At a calendar age of 11$^{5}/_{12}$ years, his carpal bones (maturity index 7) are less developed then his long (8$^{10}/_{12}$) or short bones (maturity index 8), indicating a greater defect in cuboid bones than the defect in long bones. The defective osteogenesis is also evident from the thick layer of calcified cartilage at the radius and irregular metaphysis of the ulna.

Likewise, the skeletal maturity film of a 3-year-old patient with GH insensitivity syndrome **(figure 36)** shows that carpal maturity is more severely retarded than long or short bone maturity is. Children with idiopathic short stature show uniform delay in their skeletal maturation. It is the final step of osteogenesis at the growth plate that would ultimately fuse the epiphyses with the metaphyses of the femur and tibia to a conclusion of growth. Indeed, GH therapy of children with GH deficiency accelerates bone maturation both directly, as seen in prepubertal children, and indirectly through the dispatch and acceleration of puberty during adolescence.

Fig. 35
GH deficiency. At a calendar age of 11 years and 5 months, his carpal bones (maturity index 7) are less developed than his long (8$^{10}/_{12}$) or short bones (maturity index 8), indicating a greater defect in osteogenesis than the defect in chondroplasia. The thick radiopaque rim in the radius of calcified cartilage suggests accumulation of calcified matrix at the terminal hypertrophic layer that is hindered from transforming into bone. Accumulation of calcified matrix is also evident at the carpals and distal metacarpals (arrows). Irregular surface of the ulnar metaphysis indicates defective osteogenesis of this secondary ossification center.

Fig. 36
GH insensitivity syndrome. At a calendar age of 3 years, carpal bones, with a maturity index 1, are less developed than his long (1$^{7}/_{12}$) or short bones (maturity index 2), indicating a greater defect of osteogenesis than the defect of chondroplasia. At this young age the long bones, but not the short bones, have just finished the formation of secondary ossification centers and show for the time epiphyseal growth plates.

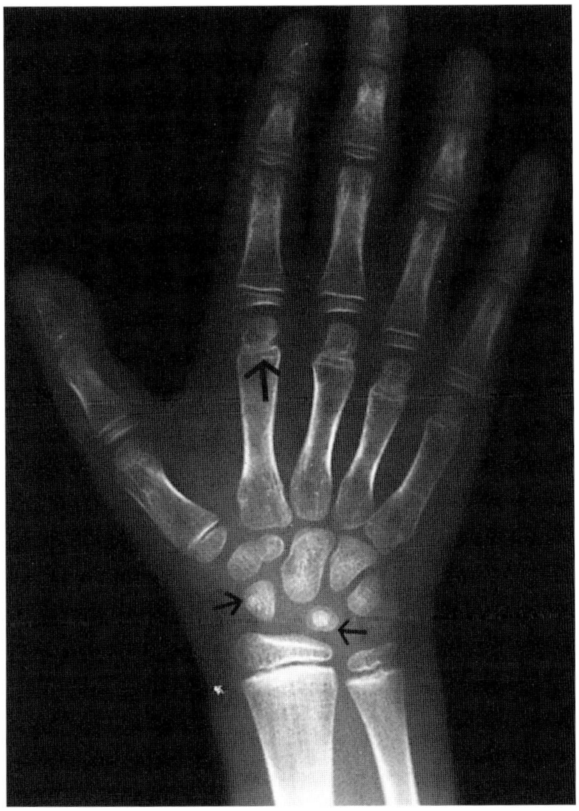

35

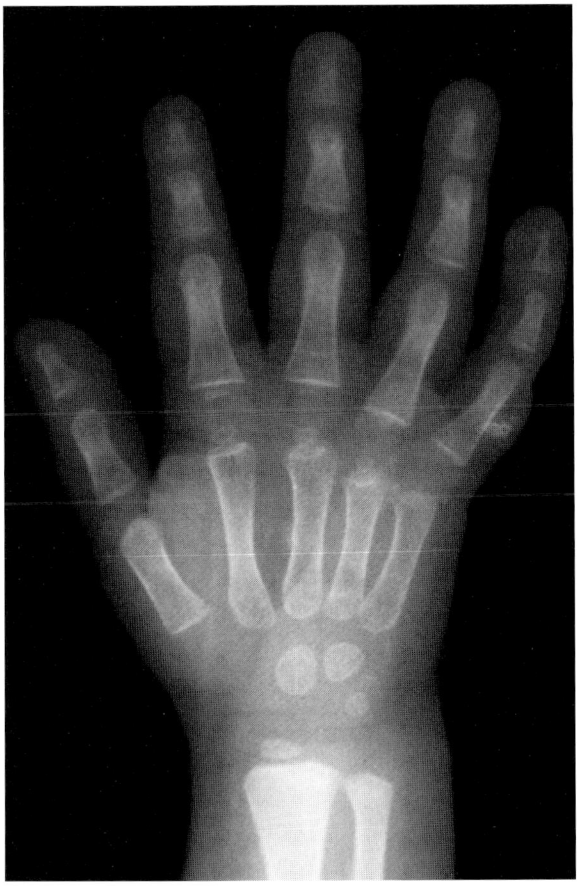

36

Thyroid Hormones
Physiological Considerations

Growth retardation and delayed skeletal maturity are foundations in the auxology of cretinism. In fact, normal thyroid hormone function is a prerequisite for any growth or bone maturation, a function which is completely arrested in severe cases of hypothyroidism. In such children, growth and bone maturation resume upon thyroid hormone replacement and catch-up growth is rapid [34]. The first description of the changes that occur during hypothyroidism in bone, cartilage and growth plate was in 1949 [35]. This report pointed out the characteristic poor growth, narrowing of the epiphyseal growth plate and disappearance of the subchondral metaphyseal trabecular bone. Induction of hypothyroidism in experimental animals induces a drastic change in epiphyseal growth plate structure and its adjacent bone [36] **(figure 37)**. The width of the epiphyseal growth plate and that of the articular cartilage are markedly diminished, as are the volumes of both epiphyseal and metaphyseal bone trabeculae. Reserve chondroprogenitor cells occupy a broader than normal segment of the growth plate. Cartilage cell columns lose their vertical organization and are jumbled. But most importantly, the growth plate is totally disengaged from its neighboring bone. The hypertrophic zone is completely sealed off from the bone marrow, and thus, osteoclasts, osteoblasts and growth factors are prevented from their task of osteogenesis. No wonder that osteogenesis comes to a standstill with a most severe delay in the radiological picture of bone maturation.

Thyroid hormones remain important regulators of bone and mineral metabolism throughout life [37]. Patients with thyrotoxicosis have increased bone turnover [38], an observation also made in patients with hypothyroidism during replacement therapy.

The mechanism underlying this drastic disarray in bone and cartilage is complex, as thyroid hormones affect such a wide range of tissues and genes [39]. One of these target tissues is the pituitary somatotroph. The GH gene was among the first to be shown to possess a thyroid response element in its noncoding promoter [40], and the secondary GH deficiency that develops during hypothyroidism is dose dependent. Indeed, clinicians have known for many years that the GH response to pharmacological stimuli is low in othyroidism and recovers upon replacement therapy. In hypothyroidism, there is also a profound decrease in GH receptors [41, 42]. But GH deficiency is certainly not the only relevant component of hypothyroidism. GH replacement therapy to hypothyroid animals accelerated growth, increased to some extent the number of proliferating cells in the growth plate, and broke down the sealing of the carti-

Fig. 37
Experimental hypothyroidism, induced by adding methimazole to the drinking water of young rats.
a Photomicrograph showing typical appearance of epiphyseal growth plate cartilage in the proximal tibia of a 14-week-old control female rat. ETB = Epiphyseal trabecular bone; HZ = hypertrophic zone; MTB = metaphysal trabecular bone; PZ = proliferative zone; RZ = reserve zone; RF = resorption front; arrow = bone marrow invasion. × 350.
b Photomicrograph showing the appearance of the epiphyseal growth plate cartilage and its adjacent tissues in the tibia of a hypothyroid female rat. The plate is narrower and the hypertrophic cells are smaller than the controls'. Cells of the reserve zone form clusters or columns that reach deep into the plate. The cartilage is sealed off from the bone marrow (arrows) while there is complete loss of the underlying metaphyseal trabecular bone. BM = Bone marrow. × 350.

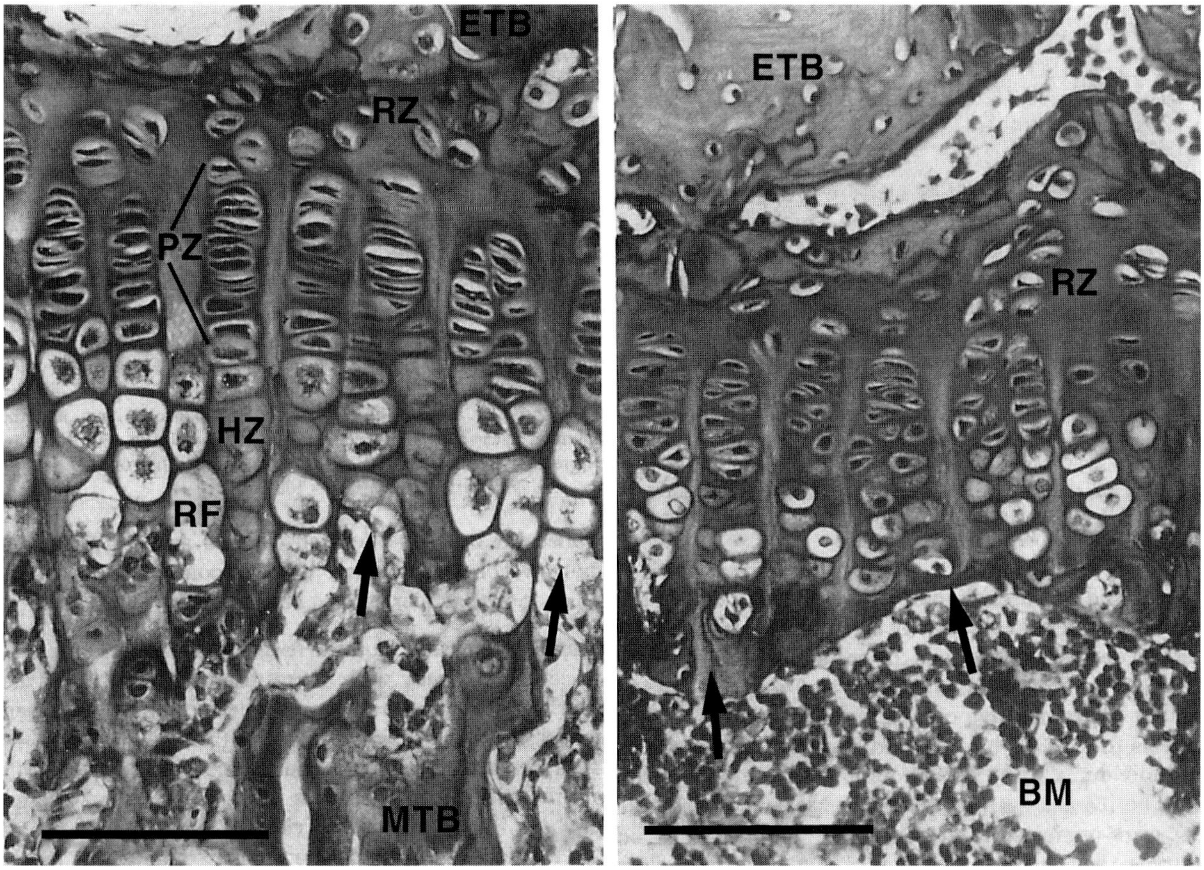

37a

37b

lage-bone interface, with no effect on cellular organization or on hypertrophic cell morphology. Yet, these affects are not as enormous as those seen after replacement therapy with thyroid hormones. Thyroid hormones exert direct effects on cartilage growth and maturation [43]. Thyroid receptors are expressed in growth plate chondrocytes and invading osteoblasts. As thyroid hormones are given to hypothyroid animals, they affect multiple specific sites at the growth plate. T3 stimulates reserve cells to differentiate into proliferating chondrocytes, it inhibits chondrocyte proliferation and stimulates hypertrophy of mature chondrocytes, as well as stimulating osteogenesis all the way from bone marrow angiogenesis at the resorption front to recovery of metaphyseal bone trabeculae. T3 alters the expression of PTHrP and PTHrP receptors. In vitro, thyroid hormones stimulate chondrocyte maturation to the hypertrophic phenotype, expressing type X collagen and alkaline phosphatase and inciting matrix calcification [44].

Implications for Skeletal Maturity Readings

Hypothyroidism in a child is associated with severe retardation of bone maturation. In severe cretinism, maturation may arrest at the neonatal phase. Radiograms of the knees give an indication of the gestational age of onset of hypothyroidism. Secondary ossification centers of the distal femur and proximal tibia epiphyses, which usually appear for the first time at gestational age 34 weeks, are absent when hypothyroidism commenced before that age. **Figure 38** shows radiograms of a femur, tibia and fibula, humerus and hand removed at autopsy from a child with cretinism who died at age 6. Ossification centers of the distal femur and proximal tibia are present but femur head, tibial distal epiphysis and capitulum of the humerus, which appear usually at age 4 months, are absent, indicating the onset of the disease between 34 weeks of gestation and postnatal age 4 months. Photomicrograph of the fibula shows complete disengagement of the cartilage from the underlying metaphysis, implying a complete arrest of osteogenesis.

Fig. 38
Hypothyroidism.
a, b Radiograms of a femur, tibia and fibula, humerus and hand removed at autopsy from a 6-year-old child with cretinism. Ossification centers of the distal femur and proximal tibia are present but femur head, tibial distal epiphysis and capitulum of the humerus, which usually appear at age 4 months, are absent, indicating the onset of the disease between gestational 34 weeks and postnatal age 4 months.
c Photomicrograph of the fibula shows complete disengagement of the cartilage from the underlying metaphysis, implying a complete arrest of osteogenesis. × 12.
d Though cartilage cells line up in columns, cores of calcified cartilage matrix do not extend into the metaphysis. Osseous tissue directly opposes the growth plate cartilage, indicating total impairment of osteogenesis. Photomicrograph. × 35.

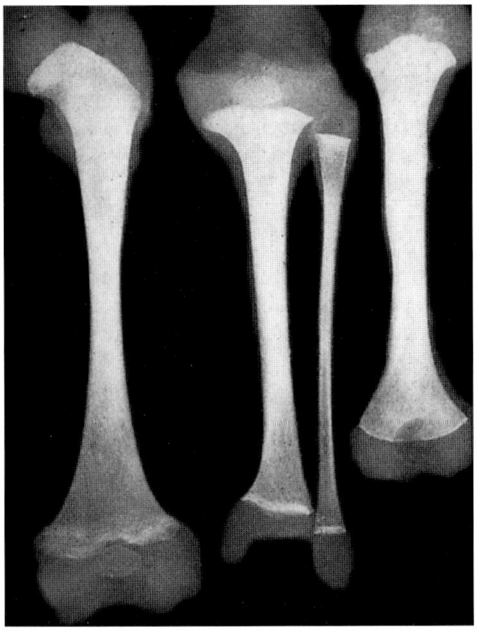

38a

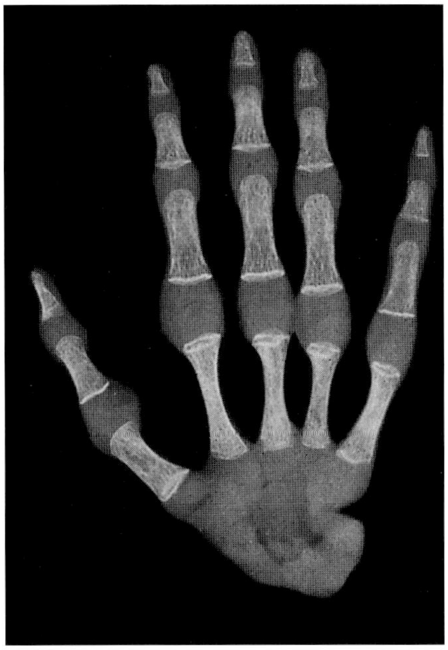

38b

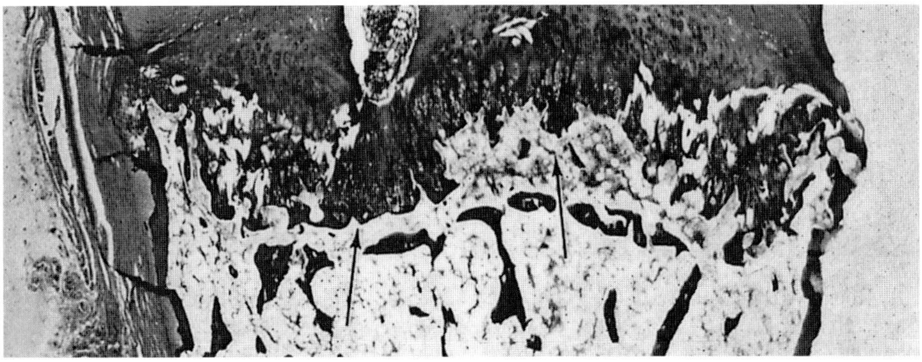

38c

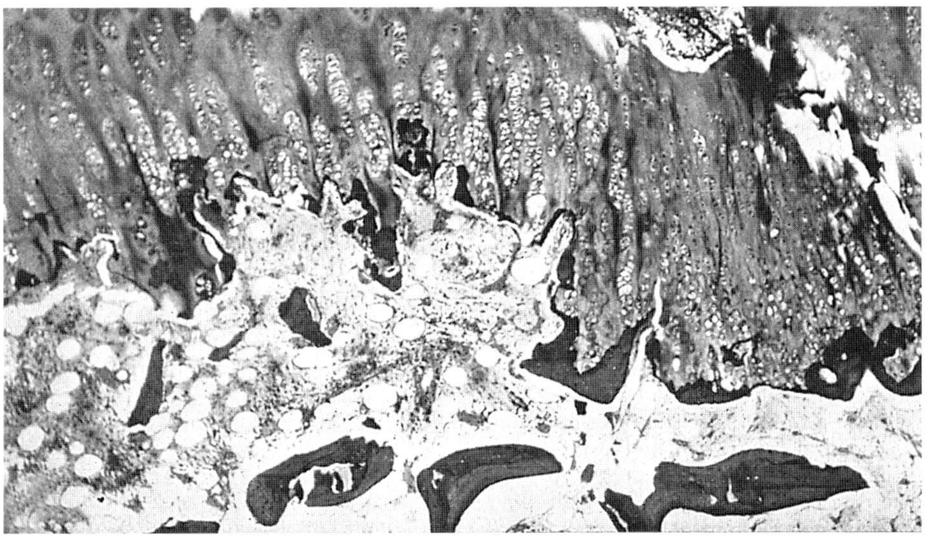

38d

If hypothyroidism remains untreated, epiphyses may remain unfused into adulthood. **Figure 39** demonstrates the findings in a 24-year-old subject with cretinism. None of the epiphyses of the phalanges or the knees are yet fused. The disengagement of cartilage and bone does not allow for fusion. The growth plate rims are sclerotic, due to excessive calcification of the cartilage matrix, but the trabecular bones are semitransparent, a sign of impaired osteogenesis.

Due to the drastic effect on osteogenesis and less severe effect on chondroplasia, patients with hypothyroidism may present with a radiological picture similar to that seen in epiphyseal dysplasia **(figure 40)**. **Figure 41** presents another aspect of epiphyseal dysgenesis in a patient with long-standing untreated hypothyroidism. Unexpected at that age, irregular islands of enchondral ossification are seen below the articular cartilage. The adjacent cartilage has undergone degeneration in some places. These islands are represented in the inset radiogram by small round radiolucent areas with radiopaque borders in the femur head, but also in the carpal bones.

After initiation of therapy in long-standing cretinism, accessory ossification centers may develop **(figure 6)**. Whereas the mechanism of this phenomenon is not understood, it may be viewed as an attempt by the bone to utilize all possible resources, even if illegitimate, for compensation after a long period of growth arrest.

Fig. 39
Cretinism in a 24-year-old male. Radiograms of part of the hand and knee show that none of the epiphyses have yet fused. The growth plate rims are sclerotic (arrow), indicating excessive calcification of cartilage matrix. Insufficient osteogenesis results in semitransparent trabecular bones.

Fig. 40
Hypothyroidism in an 8-year-old girl. The irregular epiphysis of this patient was misread as an 'epiphyseal dysplasia'. The markedly calcified cartilage matrix and the striated metaphyseal ends of both femur and tibia are due to failure of the cartilage-bone interface. Indeed, an integral part of hypothyroidism is the bone and cartilage disease, which, by radiological criteria, is an 'epiphyseal dysplasia'.

Fig. 41
Hypothyroidism.
a Radiogram of a 29-year-old male with severe untreated congenital hypothyroidism, showing the changes of 'epiphyseal dysgenesis' in the femoral heads.
b The same patient as in **a** at the age of 39 years, after receiving intermittent thyroxine replacement therapy. Resection angulation osteotomy was performed on the hip when a severe disabling osteoarthritis developed. Photomicrograph. × 7. Inset shows the radiogram just before surgery. Irregular islands of enchondral ossification are seen below the articular cartilage. These are apparent in the radiogram as small, round, radiolucent areas. The cartilage delimiting these islands has undergone degeneration in some places.

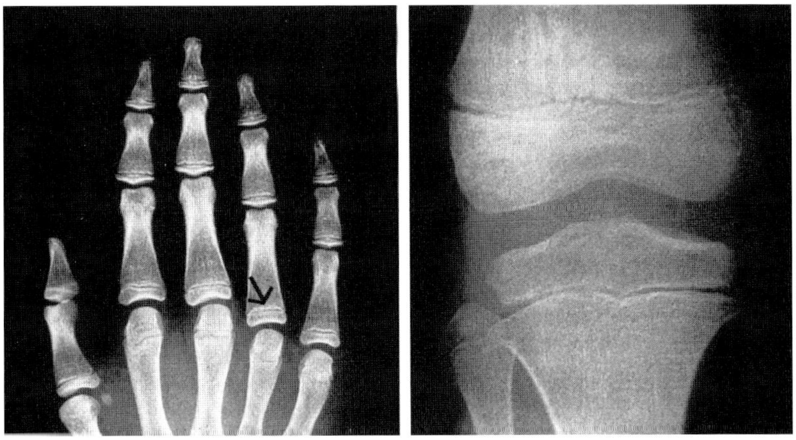

39

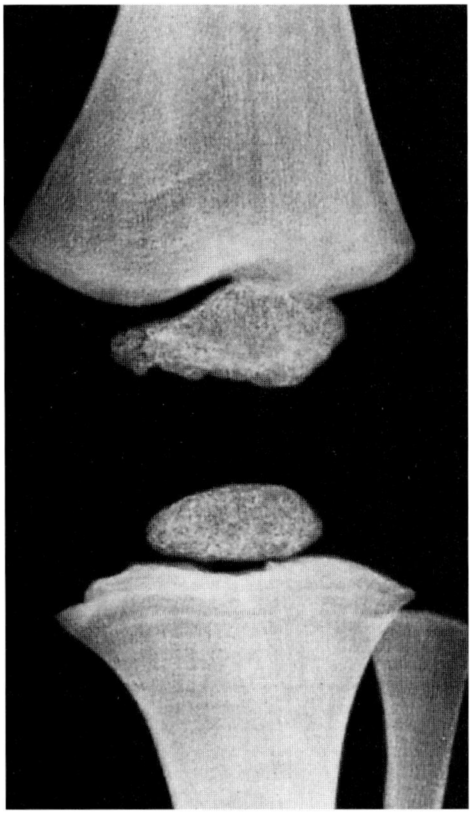

40

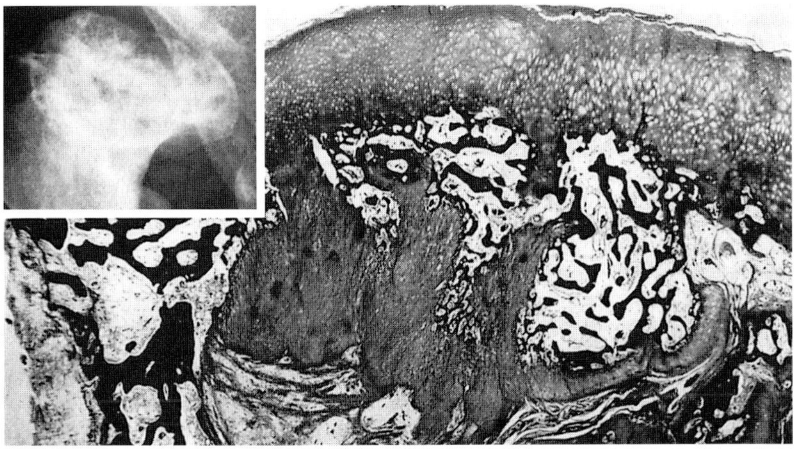

41a

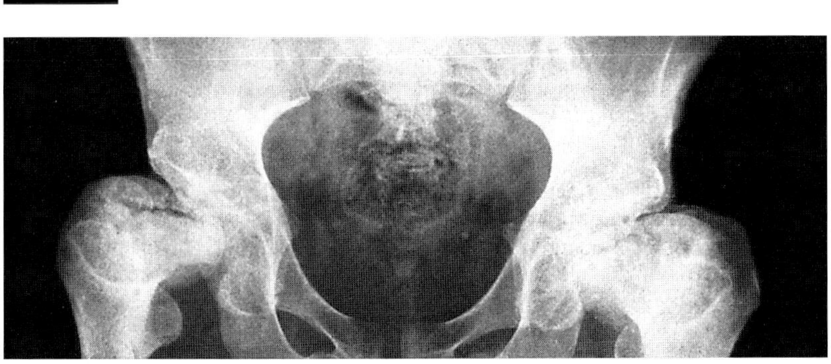

41b

Some preadolescent patients with hypothyroidism develop precocious puberty and some adolescent patients have fast puberty. Bone maturation endures then two opposing forces: sex steroids accelerate and hypothyroidism decelerates maturation. As expected, hypothyroidism prevails. When osteogenesis is hindered as forcefully as it happens in hypothyroidism, sex steroids cannot exert their stimulatory effect **(figure 42)**.

Hyperthyroidism of childhood is usually diagnosed within a short period of time after its onset, and does not show radiological signs. In a unique case of infantile Grave's disease **(figure 43)**, this female had had full-blown disease since the age of 6 months. Even at this young age, the thin cortex of the radius suggests excessive bone resorption. Seldom do we see the enhanced bone resorption seen in prolonged hyperthyroidism **(figure 44)** [45]. The thyrotoxic bone shows loss of fine granulation and longitudinal striation due to erratic rapid bone turnover. When such an effect endures for a long period of time, it develops into 'thyroid acropacy' **(figure 45)**. A milder picture of the same problem as well as osteoporosis can be seen in patients with hypothyroidism on overdose replacement therapy.

Fig. 42
Hypothyroidism with fast puberty. This 15 years and 9 months old male presented with severe hypothyroidism. Puberty advanced from Tanner stage 1 to 5 within 18 months. Selective reading of the skeletal maturity shows that his long bones and carpals had a maturity index of 12.5 and that of the short bones had a maturity index of 12. Note the irregular surface of the radial epiphysis and metaphysis, the concavity of the ulnar metaphysis and the very thin ulnar epiphysis.

Fig. 43
Infantile Grave's disease. With onset at 6 months of age, this 1-year-old infant presented with full-blown Grave's disease, confirmed by thyroid-stimulating immunoglobulins. The skeletal maturity is far advanced for her age. Selective reading of this film shows the maturity index of the long bones to be 2.5, the short bones 3.5 and the carpals 3. Even at this young age, the thin cortex of the radius suggests the bone-resorbing effect of thyroid hormones.

Fig. 44
Thyrotoxicosis. Compared to the normal metacarpal **(a)**, the thyrotoxic bone **(b)** shows loss of fine granulation and longitudinal striation due to erratic rapid bone turnover. The arrows indicate a nutrient vessel.

Fig. 45
Thyroid acropacy. This patient had Grave's disease and thyroidectomy 11 years previously and the bone disease has persisted throughout these years. Other than the soft tissue swelling of the 2nd and 4th fingers of the left hand, thickened cortex and periosteal reaction is noted in the short bones.

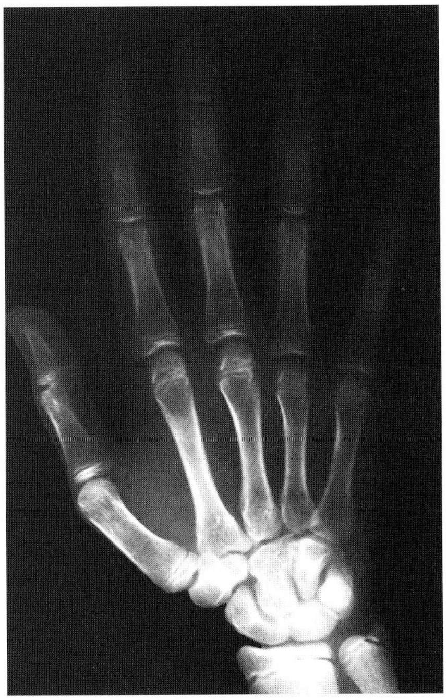

42

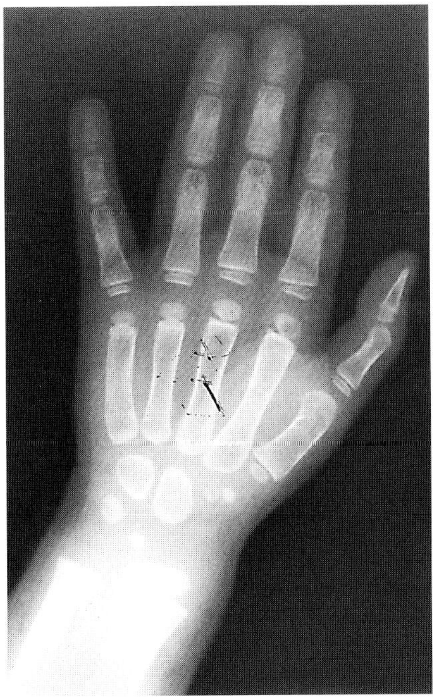

43

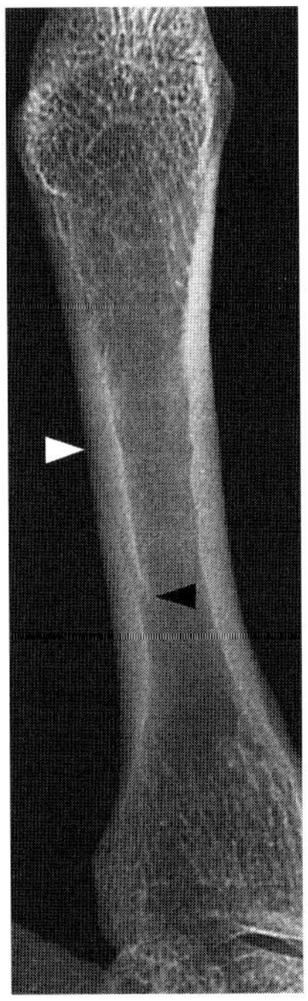

44a

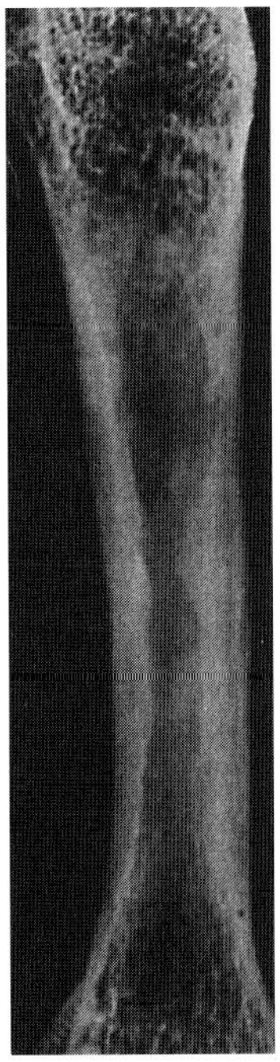

44b

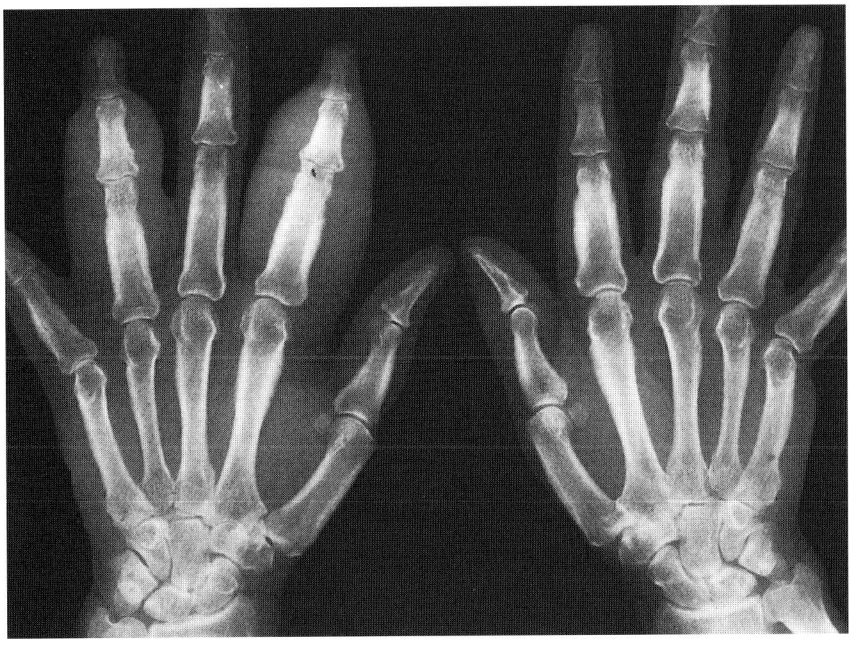

45

Sex Steroids (Gonadal and Adrenal)

Physiological Considerations

Sex steroids of gonadal origin are essential for the growth spurt and bone maturation at puberty, as well as for bone mineral homeostasis, as is evident from studies in postmenopausal osteoporosis. The role of adrenal androgens is less well defined. Some of their effects may be indirect, through estrogen action on the GH-IGF-1 axis, but others are direct effects on cartilage and bone. Estrogen and progesterone receptors are abundant in proliferating and hypertrophic chondrocytes as well as in osteoclasts, where estrogens inhibit bone resorption and osteoblast activity [46] and alter the release of cytokines [47]. Thus, oophorectomy results in wider proliferative and hypertrophic zones, apparently as a consequence of osteogenesis arrest. Estrogens increase extracellular matrix generation and the concentration of matrix growth factors. Some estrogen effects may be due to its role in PTH/PTHrP activity. One example is the reduced fibronectin production by PTH seen in humans and rats with estrogen deficiency. The many conflicting results of estrogen effects on bone are now better understood with the identification of two classes of estrogen receptors, with their tissue specificity. Thus, tamoxifen is an anti-estrogenic agent in the breast, with partial agonistic effects on bone. Knockout experiments of estrogen receptors and the discovery of a human mutation in the estrogen receptor taught us a lesson on the crucial role of estrogen in bone maturation. An abundance of data on the role of estrogen in postmenopausal bone loss has direct implications to the field of bone maturation during childhood and adolescence.

Androgens may be just as important in women as in men [48]. While some of the androgenic effects may be induced through aromatization into estrogen, bone cells and epiphyseal chondrocytes express androgen receptors [49] and respond to androgen both in vitro and in vivo [50]. Human osteoblast-like cells possess androgen receptors and respond to androgen by proliferation, differentiation, mineralization and expression of genes consistent with bone formation. Both during puberty and aging, administration of androgens has positive effects on mineralization [51] and bone mass. These may be direct effects on bone or indirect effects through the anabolic effects the androgens have on muscle.

Hypogonadal men develop unique body proportions, recognized as eunuchs **(figure 46)**. The extremities are disproportionately long in relation to the trunk, due to late closure of the epiphyseal growth plates, long after the cuboid vertebrae had ossified and calcified. Interestingly enough, longitudinal growth continues despite the delay in bone maturation, giving emphasis to the distinct mechanism of each of these processes. In a mirror image of that seen in Turner syndrome, increased growth is evident from the distal to the proximal and from the lateral to the medial bones.

Fig. 46
Eunuch. This 28-year-old man was 6 feet tall. The extremities (long bones) are disproportionately long in relation to the trunk. Distal phalanges are longer than proximal phalanges. The femur heads and the iliac crests have not yet fused.

46

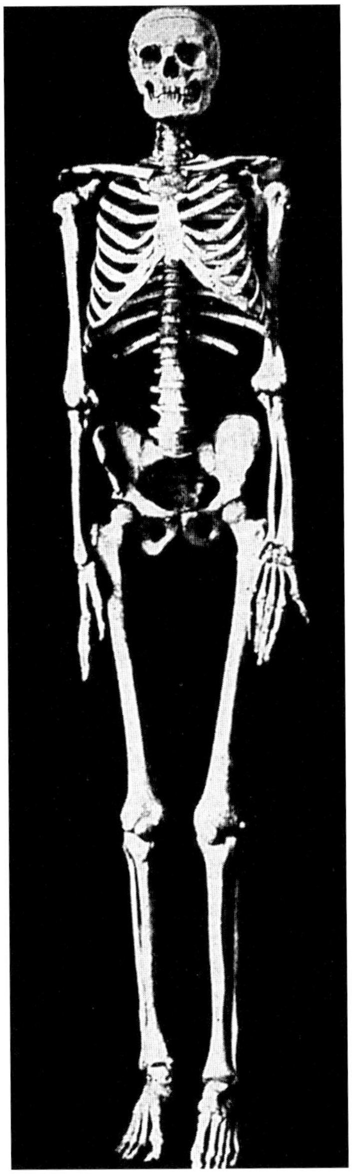

Delayed and Precocious Puberty

The fundamental role of sex steroids in bone maturation is obvious from the marked delay in skeletal maturity of children with delayed puberty and its advancement in children with precocious puberty. So great is the effect of sex steroids, that the tempo of bone maturation almost unanimously loses any relationship to a child's calendar age during adolescence. Thus, the distance between a female skeletal maturity index of 10 and 11 is far greater then it is in a prepubertal index of 6 – 7 and the tempo of bone maturation relates better with pubertal development than it does with calendar age **(figure 47)**.

Bone maturation in precocious puberty is characterized by advanced maturity of both chondroplasia and osteogenesis, but the former is more profoundly affected. Thus, the film in **figure 48** of this 5 years and 8 months old boy who suffered an optic glioma and central precocious puberty shows the short bones to have a maturity index of 13 whereas the long bones and the carpals show an index of 10. The film in **figure 49** of a male aged 14 years and 8 months, with constitutional delay of growth and puberty, shows the short bones to have a maturity index of 11 whereas the long bones and the carpals show an index of 12.

Fig. 47
Bone maturation during puberty. The tempo of bone maturation relates better with pubertal development than it does with calendar age. In T.E. who started puberty (Tanner B2) at age 13, bones matured more slowly (bone age/calendar age advancement <1) than it did in W.Z. who started puberty at age 7 and a half (bone age/calendar age advancement >1).

Fig. 48
Precocious puberty. This 5 years and 8 months old boy had an optic glioma and central precocious puberty (Tanner stage 4). The short bones are far more advanced, with a maturity index of 13, than the long bones and the carpals, which show an index of 10. Maturation of the short bones ahead of the carpals indicates that chondroplasia was affected by premature sex steroids secretion more than chondroplasia was.

Fig. 49
Delayed puberty. This 14 years and 8 months old male had constitutional delay of growth and puberty (Tanner stage 2). His short bones have a maturity index of 11 whereas the long bones and the carpals show an index of 12. The greater delay in maturation of the short bones reflects the role of sex steroids deficiency in chondroplasia. It is felt, though unproven, that the maturity index of the short bones mirrors the maturation of the legs' long bones.

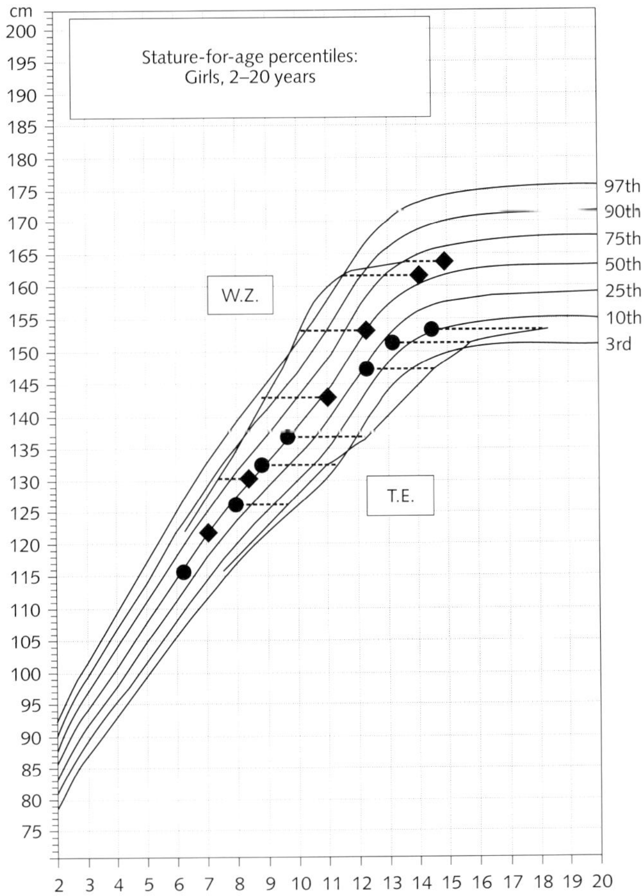
Stature-for-age percentiles:
Girls, 2–20 years
cm
200
195
190
185
180
175
170
165
160
155
150
145
140
135
130
125
120
115
110
105
100
95
90
85
80
75
2 3 4 5 6 7 8 9 10 11 12 13 14 15 16 17 18 19 20
97th
90th
75th
50th
25th
10th
3rd
W.Z.
T.E.

47

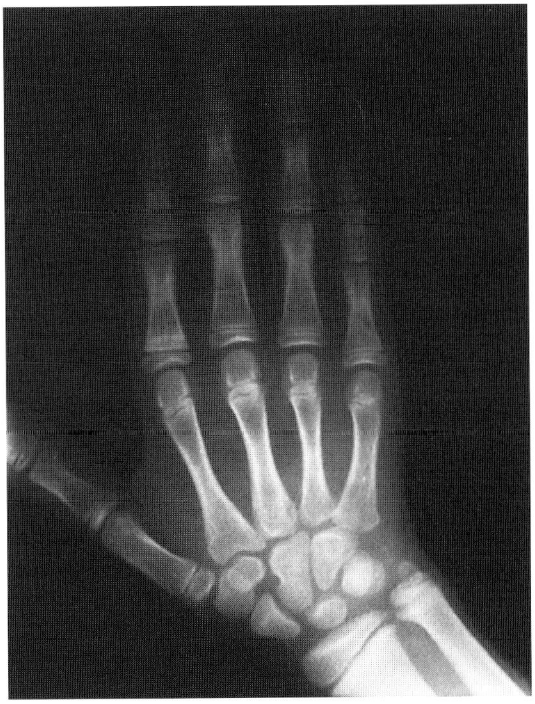

48

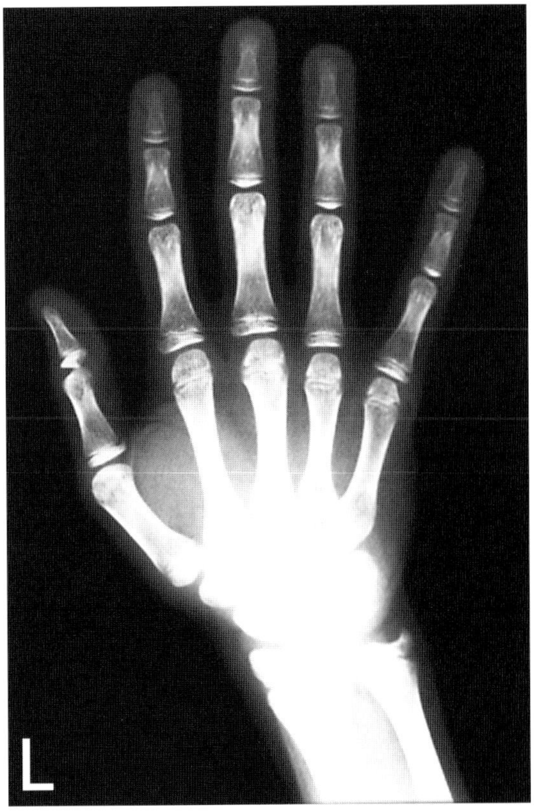
L

49

But even before puberty, adrenal androgens and ovarian estrogens exert specific effects on bone maturation, as inferred from several clinical conditions. From early infancy on, girls' bones mature more rapidly than do those of boys. The secondary ossification center of the radial end is apparent on X-ray films of girls at 6–10 months, but would appear in boys at 12–24 months. Reading a series of X-ray films according to the standards of reference for both male and female hands and wrists reveals a discrepancy that evolves in two phases **(figure 50)**. During the first 3 years of life, girls precede boys' bone maturity index by as much as 1. It is assumed that this is an effect of low-level estrogen production by the infantile ovary. After quiescence in prepuberty, pubertal bone maturation of girls advances faster to precede the index of boys by 3 by the end of girls' puberty at age 14 years.

The role of adrenal androgens in bone maturity can be appraised from the film of a child with adrenal glucocorticoid and androgen deficiency. The child, whose hand is depicted in **figure 51**, had familial ACTH unresponsiveness. Such children have selective deficiency of adrenal glucocorticoids and androgen, with intact mineralocorticoid content. At age 14 years and Tanner stage 3, his carpals, with a maturity index of 9.5, are more retarded than his long and short bones, with a maturity index of 10.5.

The role of adrenal androgens is well established for its rapid maturational impact on bone development in children with congenital adrenal hyperplasia. The effect is greatest on the short bones, intermediate on the cuboid bones and minimal on the long bones of the hand **(figure 51b)**. For a not completely understood reason, this effect is not exerted before birth or within the first 6 months of life. As adrenarche develops, the serum levels of adrenal androgen increase and accelerate bone maturation.

During GnRH analogue therapy to girls with precocious puberty, when gonadal sex steroids are suppressed, adrenarche is associated with advancement of bone maturation [52]. On the other hand, administration of adrenal androgen to children with delayed or absent adrenarche does not accelerate bone maturation.

Fig. 50
Sexual dimorphism. The reading of both male and female bone maturity in a series of radiograms extending from 3 months through 19 years **(a)** reveals the gap between the sexes **(b)** which stems from the faster maturation of the female skeleton. Whereas the X and Y chromosomes may have other unknown genes that affect bone maturation, for simplicity of the discussion, one can assume that the gap is an effect of sex steroids. It is evident that the gap starts to form over the initial 3 years of life, plateaus during prepuberty and increase again as females undergo puberty before boys do.

Fig. 51
Adrenal androgens.
a At age 14 years and Tanner stage 3, this child with familial ACTH unresponsiveness, glucocorticoid and adrenal androgen deficiency had a carpals maturity index of 9.5 and a long and short bones maturity index of 10.5.
b This 1 year and 1 month old boy with congenital adrenal hyperplasia has a maturity index of 1 for his radius and ulna, and index of $1^4/_{12}$ for his carpal bones and 2 for his short bones of the hand.

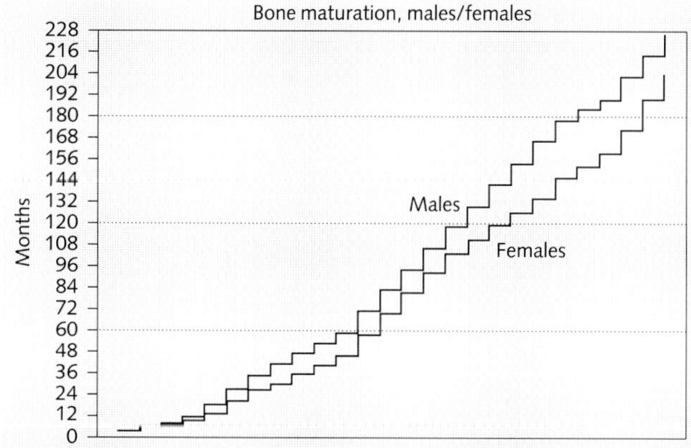

50a

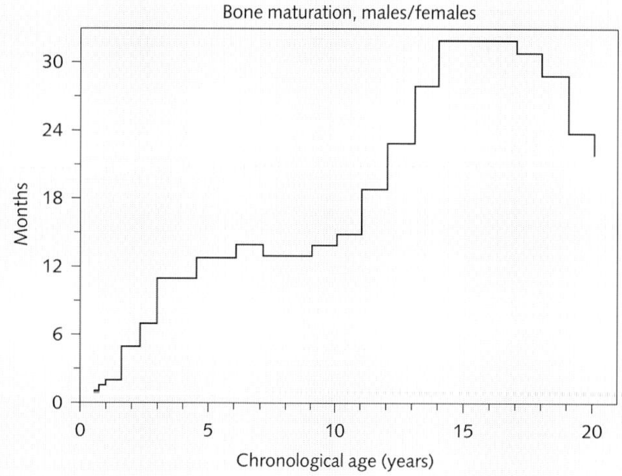

50b

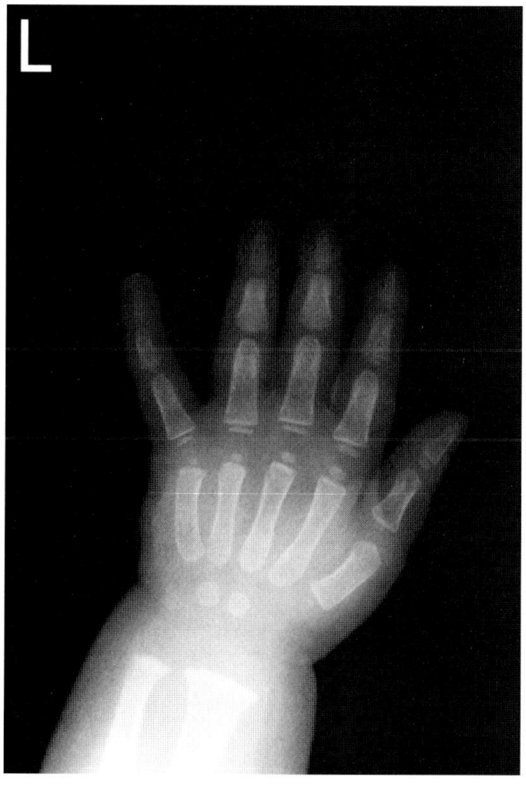

51a

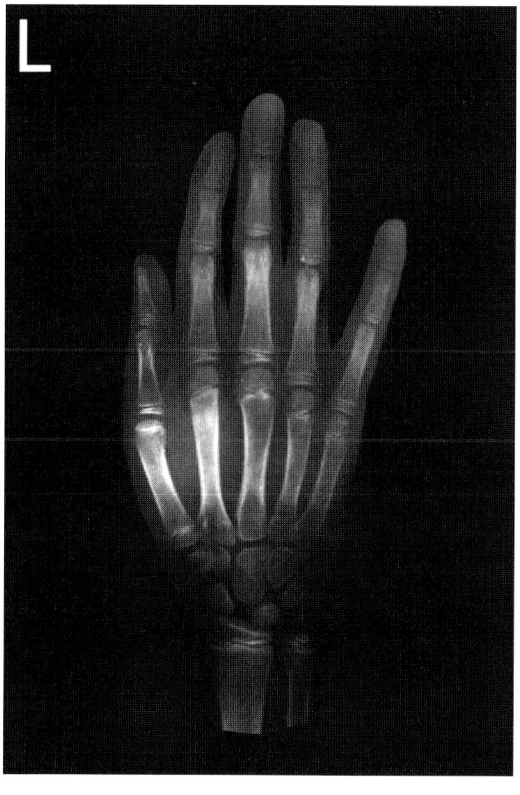

51b

Another clinical condition of great interest is patients with loss-of-function mutations of the estrogen receptor **(figure 52)** [53] or androgen-aromatase deficiency [54]. In these males who either lack estrogen or have estrogen insensitivity, bone maturation is much delayed. The radius ulna and short bone indices are uniformly 15 and the cuboid carpal bones are fully matured (maturity index 18). Regions of osteogenesis at the carpals and growth plates are sclerotic.

Fig. 52
Estrogen effect. Estrogen insensitivity in a 27-year-old man on the right, compared to a normal film of a 15-year-old subject.

Turner Syndrome

A special case of interest with regard to the impact of estrogen on bone maturation is girls with Turner syndrome. To identify the short 4th metacarpal, one does not need an X-ray film **(figure 53)**. Other than their size, the short metacarpals show large epiphyses that are prematurely fused to the metaphyses.

The mechanism of growth retardation in Turner syndrome has not been resolved. The combination of short stature, irregular metaphyses, vertebral anomalies and Madelung's deformity [55, 56], lead to the assumption that Turner's syndrome is a bone dysplasia. The Madelung's deformity **(figure 54)** is also apparent in Leri-Weill syndrome and, indeed, a deficiency of the X chromosome product, SHOX protein, has been implicated in both these syndromes. The unique defect in the medial aspect of the radius and sometimes the ulna as well suggests that SHOX protein influences the marginal regions more than it affects the central parts of the growth plate.

Fig. 53
Turner syndrome in an 18-year-old young woman. Skeletal maturation is markedly retarded when estrogen replacement therapy is delayed. The phalanges are more delayed than are the carpal bones. The short 4th and 5th metacarpals are clinically evident and result from premature fusion of the enlarged epiphysis to the metaphyses.

Fig. 54
Madelung's deformity in a 12-year-old girls with Leri-Weill syndrome. The radius is bowed laterally, tipping its distal end toward the ulna, leaving a V-shape between them.

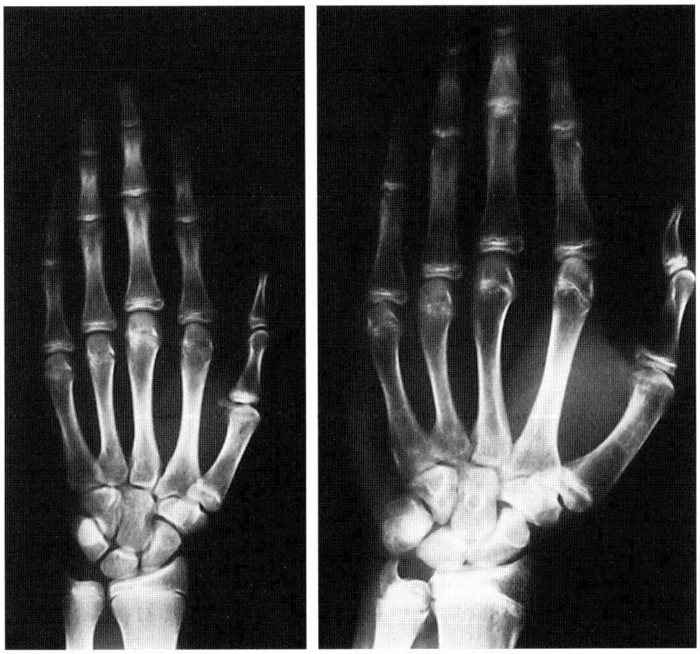

52

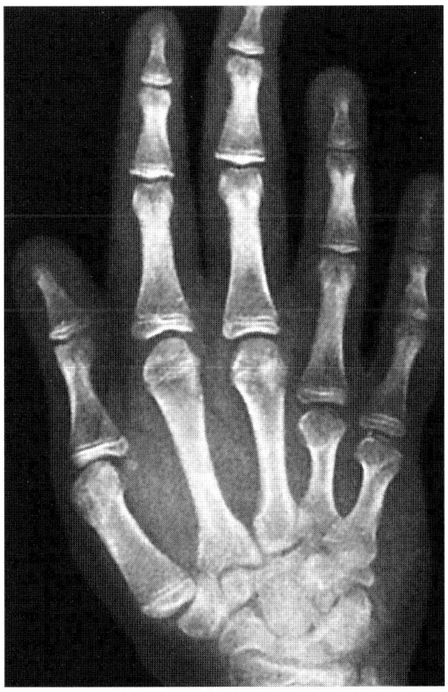

53

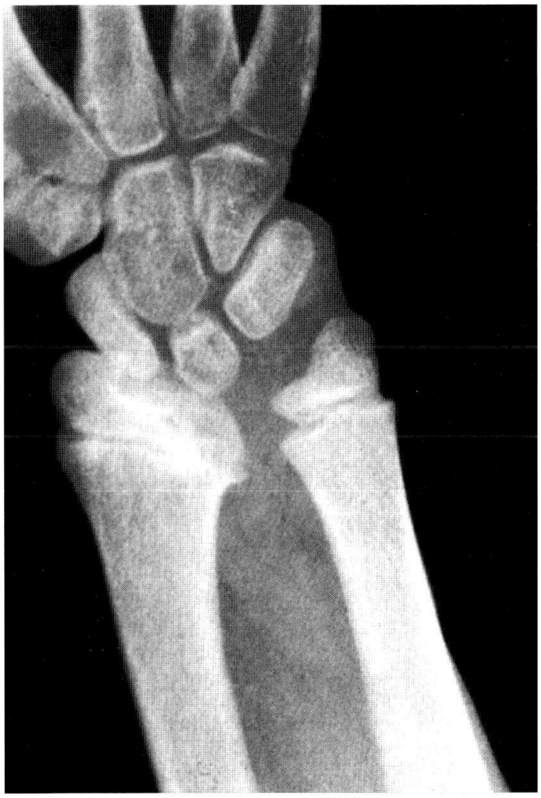

54

A delay in bone maturation is seen in both prepubertal and pubertal Turner girls [57]. Metacarpal and phalangeal bones are shorter and narrower in Turner patients, with increasing growth deficiency from distal to proximal and from lateral to medial bones [58, 59]. During prepuberty, Turner patients show only a mild delay in skeletal maturation, with the greatest delay in maturation of the short bones and least delay in the radius and ulna (long bones) and metacarpals. The carpal bones show intermediate retardation. During adolescence, the phalangeal bones are further retarded and the carpal bones show a moderate retardation **(figure 55)** [60]. The unique profile of bone maturation in Turner syndrome suggests an insult to chondroplasia, which may be related to estrogen deficiency or to an as yet undetermined endocrine or paracrine derangement. A photomicrograph of the humerus growth plate of a 32-year-old woman with gonadal dysgenesis shows an unfused plate, with disorganized chondroplasia and apparently normal osteogenesis **(figure 56)** [4].

Fig. 55
Turner syndrome. Third-degree regression of the delay in bone maturation against chronological age in patients with Turner syndrome. Whereas in idiopathic short stature long bones, short bones and carpals show parallel regressions with narrow gaps, the pattern in the Turner patients showed a milder delay in the 'short bones' during early prepuberty, with increased delay during adolescent years. The carpals and 'long bones' were relatively more retarded in prepuberty, and less so in adolescence.

Fig. 56
Turner syndrome. A photomicrograph (×15) of a humerus growth plate of a 32-year-old woman with gonadal dysgenesis, showing an unfused plate, with disorganized chondroplasia and apparently normal osteogenesis. The bone trabeculae are extraordinary thin.

55

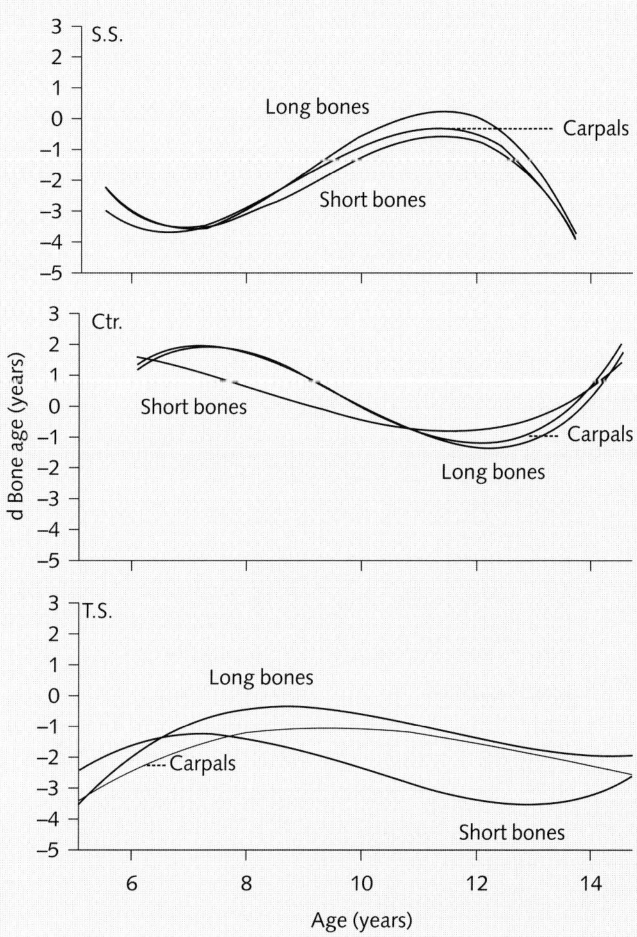
S.S.
Long bones
Carpals
Short bones
Ctr.
Short bones
Carpals
Long bones
T.S.
Long bones
Carpals
Short bones
d Bone age (years)
Age (years)
3
2
1
0
–1
–2
–3
–4
–5
6
8
10
12
14

56

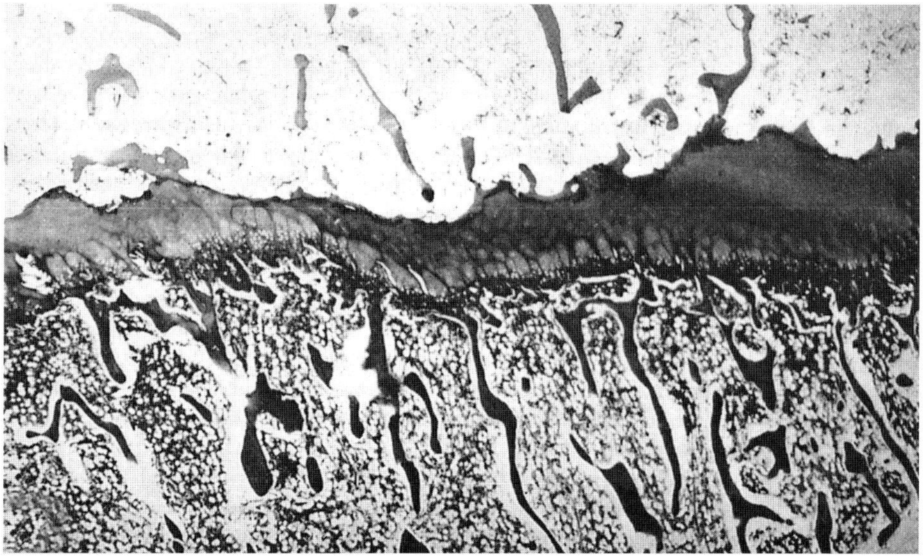

Multiple Hormone Deficiency

Combined hormonal deficiencies show more profound defects in bone maturation and enable further characterization of the hormonal impact on bone maturation. The 22-year-old man in **figure 57** has an X-linked hypogonadotropic hypogonadism and adrenal insufficiency due to a DAX1 defect. With deficiency of both his adrenal and testicular androgens, he has developed an extreme delay in bone maturation.

Figure 58 shows another extreme case of delayed bone maturation. At 39 years of age, this untreated patient with panhypopituitarism has an arrested osteogenesis, as evident from the picture of the metaphyseal plate. An area of irregular calcification is immediately followed by longitudinal trabecular formation. The growth plate is unable to fuse due to sealing of the calcified cartilage from the adjacent bone marrow. A combination of hypothyroidism, hypogonadism and GH deficiency renders this physis so immature. A photomicrograph of a distal femur end from a similar patient is seen in **figure 59** [4]. The plate is irregularly wavy with focal areas of degeneration. A complete arrest of osteogenesis at the cartilage-metaphyseal interface is due to a combined deficiency of TSH and GH.

Fig. 57
Hypogonadotropic hypogonadism and adrenal insufficiency in a 22-year-old man with DAX1 defect. With deficiency of both his adrenal and testicular androgens, he has developed into an extreme delay in bone maturation. The maturity index of his short and long bones is 14 and that of the carpal bones is 15, indicating the effect of sex steroids on the process of chondroplasia more than on osteogenesis.

Fig. 58
Panhypopituitary.
a An 18-year-old male with multiple pituitary hormone deficiency (GH, TSH, prolactin, LH and FSH deficiency). The long and short bones show an index of 7 and the carpals have an index of 6.
b A 39-year-old untreated patient with hypopituitarism. An arrest of osteogenesis is evident from the picture of the metaphyseal plate. An area of irregular calcification is immediately followed by longitudinal trabecular formation. The growth plate is unable to fuse due to sealing of the calcified cartilage from the adjacent bone marrow. A combination of hypothyroidism, hypogonadism and GH deficiency renders this physis so immature.

Fig. 59
Hypopituitarism.
a Photomicrograph of a distal femur end from a 38-year-old patient. The plate is irregularly wavy, with focal areas of degeneration. × 7.
b Greater detail of **a**. A complete arrest of osteogenesis at the cartilage-metaphyseal interface is due to a combined deficiency of TSH and GH. × 27.
c Photomicrograph of the costochondral junction from the same patient. Cartilage cells are lined up in short columns, an arrangement which is clearly abnormal for an adult. × 22.

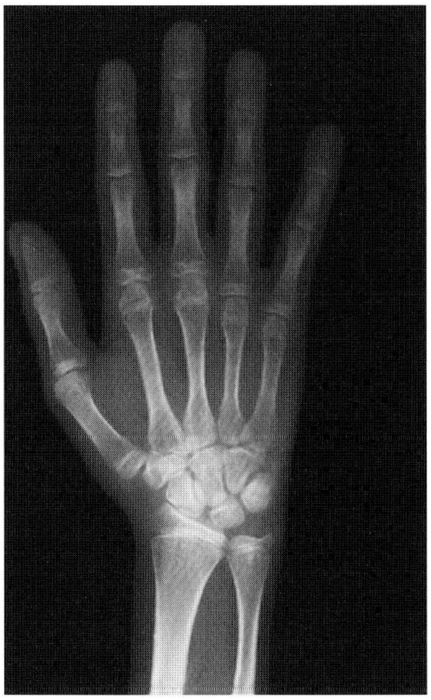

57

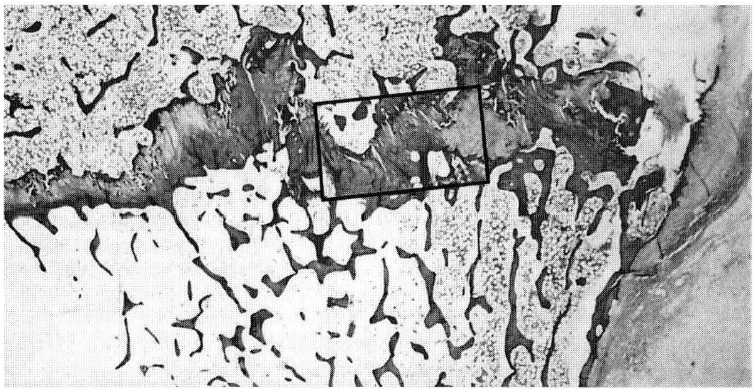

59a

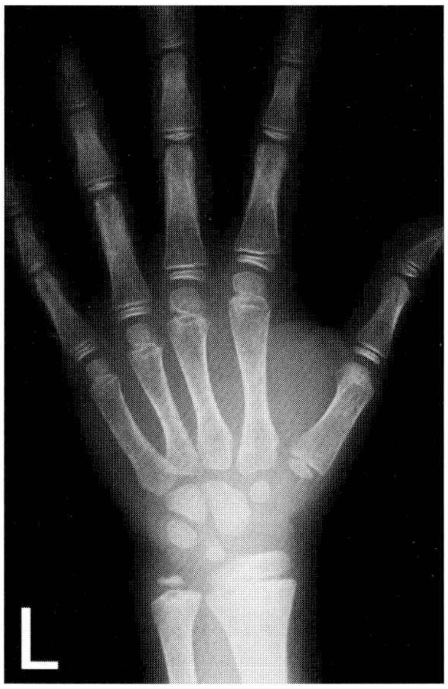

58a

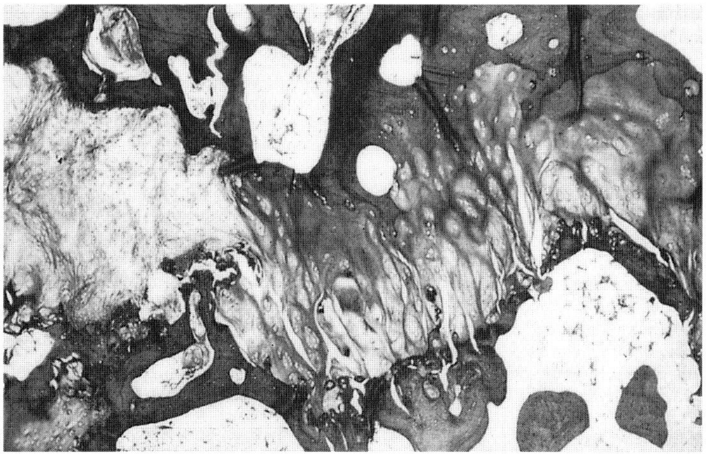

59b

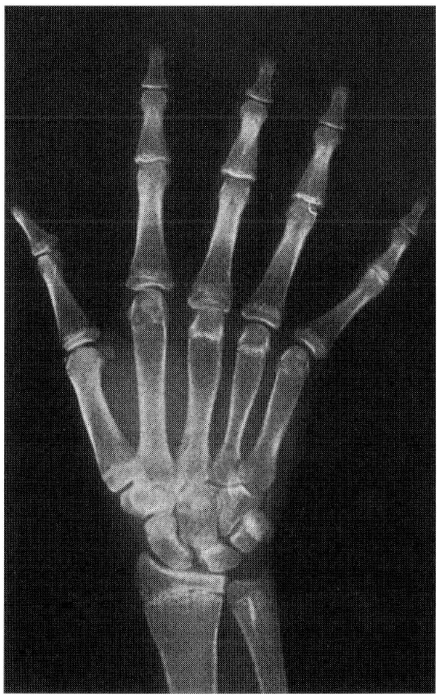

58b

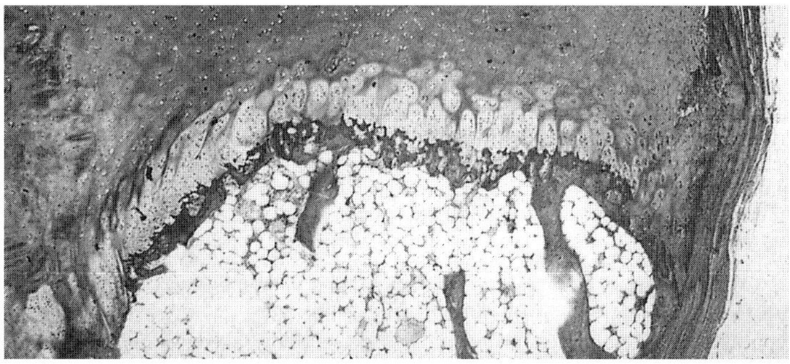

59c

Implications for Skeletal Maturity Readings

Skeletal maturity index does not have any better relationship to the onset of puberty or the initiation of pubertal growth spurt than does chronological age. Yet, in patients with delayed puberty it correlates better than chronological age with the onset of secondary sexual characteristics. It is also a fairly good predictor of the age of menarche in girls and the age of peak growth velocity in boys.

The original Greulich and Pyle atlas of bone development provided separate sets of figures for males and females, but several subsequent editions attempted to combine them into a single set of figures. As androgens and estrogens have a different impact on biological variables in bone maturity, and despite the fact that some of the male's bone maturation is estrogen-dependent, it makes little sense to read female and male maturity indices from the same set of atlas figures.

More than any other age group, adolescence is the period where individual bone indices give much more information than do their sum or their average. Sex steroids affect the short bones of the fingers and metacarpals more than the cuboid carpals or even the radius and ulna. With the physiological deceleration of growth during the immediate prepubertal years, which may get to extremely slow growth in children with delayed puberty, it is often difficult to scrutinize pathological growth arrest from physiological deceleration. A delay in the short bone index is more likely to be physiological than a delay in the carpal bones that characterizes GH deficiency.

Sex steroids do not affect bone development in utero or during the first 6 months of life. Thus, the bone maturity index of newborns with congenital adrenal hyperplasia is normal at birth and in the subsequent months. As soon as acceleration of bone maturity starts, the unique pattern of greater impact on the short bones becomes evident.

Following pathological growth aberrations and their recovery, growth catches up or catches down to 'canalize' into the original growth percentile. The mechanism of this unique phenomenon is poorly understood, but it is quite extreme after correction of puberty-related growth deviation; bone maturation plays an important role in this event. Thus, following glucocorticoid replacement therapy in congenital adrenal hyperplasia, catch-down growth is associated with complete arrest of bone maturation that can last years until canalization into the child's 'growth plan' **(figure 60)**. The same is true for patients with central precocious puberty, as they are treated with gonadotropin-suppressing GnRH analog. At the other extreme, as patients with delayed puberty are treated with sex steroids, rapid bone development consorts with a catch-up growth.

Fig. 60
Complete arrest of bone maturation in a patient with congenital adrenal hyperplasia (CAH) and a patient with central precocious puberty (CPP), as they are treated successfully to inhibit sex steroid production.

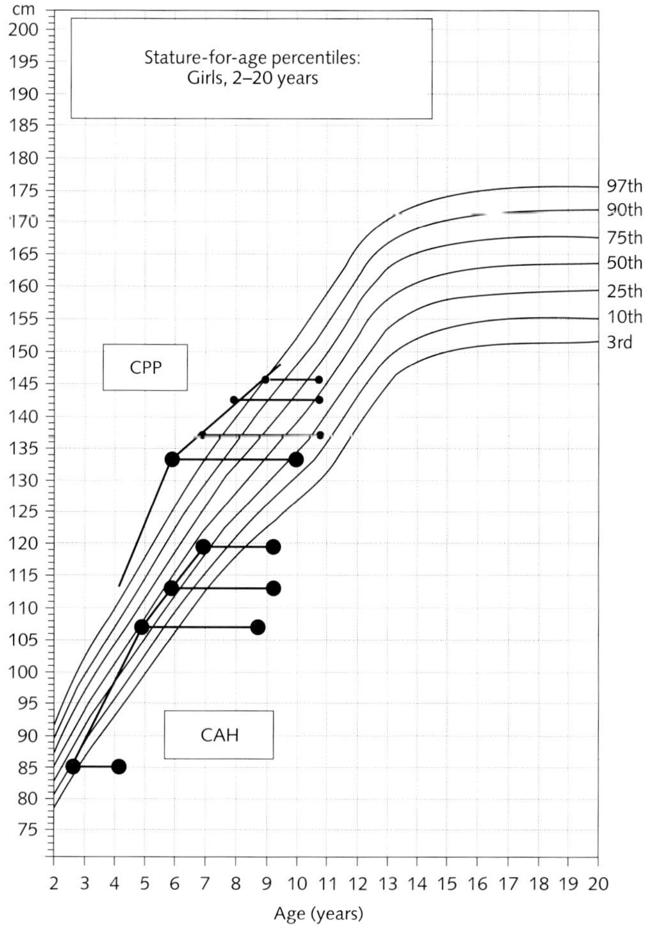
Stature-for-age percentiles:
Girls, 2–20 years
cm
CPP
CAH
97th
90th
75th
50th
25th
10th
3rd
Age (years)

Calcium-Regulating Hormones

Physiological Considerations

The most obvious of the calcium-regulating hormones to affect bone maturation is vitamin D through its active metabolite 1,25-dihydroxyvitamin D. The bone disease inflicted by vitamin D deficiency or errors in vitamin D metabolism is rickets. Rickets is a disease of the growth plate. The same metabolic condition would inflict a different type of bone disease in adults, where growth plates have been fused into a solid bone. This bone disease, osteomalacia, produces a totally different entity and a different radiological picture. Vitamin D, through its receptor, is involved in the differentiation of the growth plate chondrocytes, modulates terminal differentiation [61] and reduces alkaline phosphatase activity and type X collagen both in vitro and in vivo [62]. In the adjacent bone, the osteoblast is the primary target cell for vitamin D, where it affects the transcription of a variety of matrix proteins and enhances bone mineralization. However, vitamin D is also a major bone-resorbing hormone, through its effect on osteoclast mobilization. Whereas experimental data show this multitude of specific vitamin D effects, they all may be of secondary importance to its calcium-resorbing role. In reality, patients with rickets due to loss-of-function mutations in the vitamin D receptor and, thus, complete vitamin D insensitivity, have complete recovery when given high-dose calcium [63].

In vitro experiments show a role for PTH in the growth plate, with an effect both on resting and growing cells [64], stimulating chondrocyte proliferation and inhibiting matrix deposition and alkaline phosphatase activity [65]. To accomplish these effects, the tissue must locally supply growth factors. Whereas the in vivo effects on the growth plate itself may be small, PTH has profound effects on elements of the bone tissue that are involved in osteogenesis. Thus, PTH recruits osteoclasts from the bone marrow to initiate calcified matrix resorption. On the other hand, intermittent administration of PTH is anabolic to the bone [66] and has been utilized successfully in the treatment of osteoporosis. Mediated by IGF-1 it stimulates osteoblast differentiation from its osteoprogenitor cells to accomplish bone formation as the last phase of osteogenesis.

While both PTH and PTHrP utilize the same receptor, the physiologic role of the latter is unique. PTHrP and PTHrP receptors are expressed in osteoblasts of primary ossification centers and in maturing chondrocytes of the growth plate [67]. PTHrP regulates the proliferation of growth plate chondrocytes and enhances matrix mineralization. PTHrP is part of a cascade of regulatory factors that includes the Indian hedgehog upstream and Bcl-2, a protein that controls programmed cell death in several cell types, downstream. **Figure 61** shows the growth plate of a transgenic mouse to PTHrP. Bone maturation towards hypertrophy and programmed cell death is delayed [68].

Fig. 61
PTHrP. As compared to a control animal, a transgenic mouse for PTHrP shows clear delay in chondrocyte maturation toward hypertrophy and programmed cell death.

61

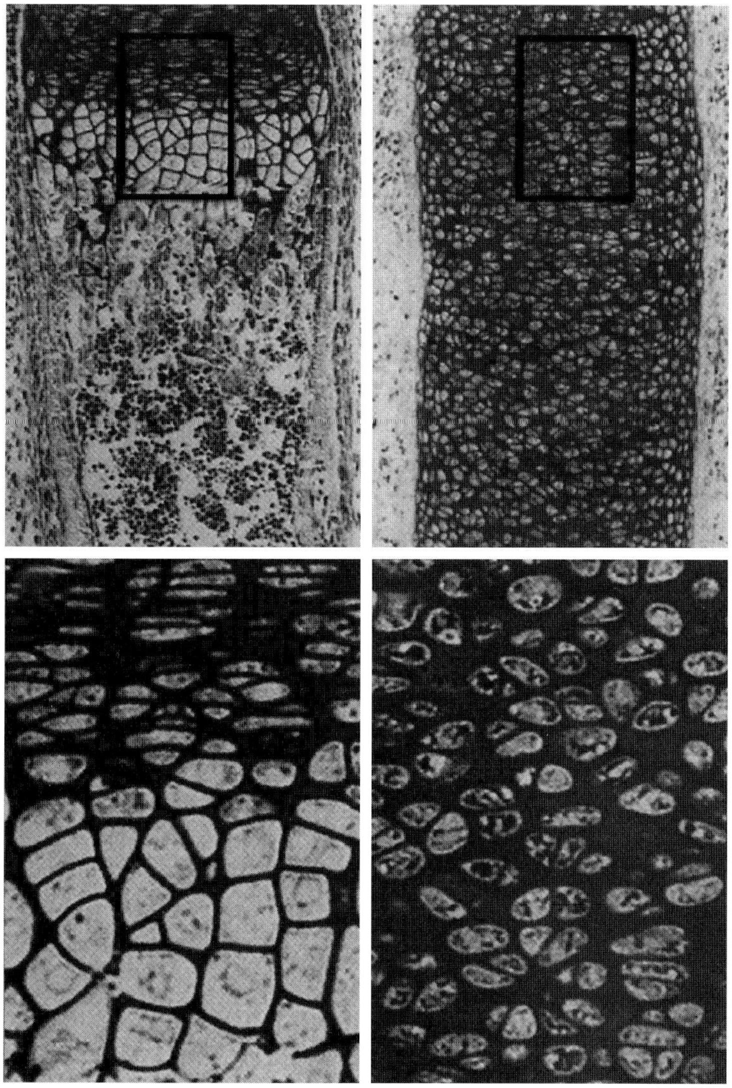

Implications for Skeletal Maturity Readings

Rickets is characterized by expansion of the growth plate margins, by an increase in hypertrophic zone growth and calcification, as well as by impaired marrow invasion [69] and bone mineralization. In severe cases, it is associated with changes induced by excess parathyroid hormone **(figure 22)**. Selective reading of the 3 bone types shows a more severe delay in the maturation of the cuboid bones (osteogenesis), as compared to the short or long bones (chondroplasia). This is illustrated in a photomicrograph of a child with active rickets who died at the age of 18 months **(figure 62)**. Widening of the wrist is evident clinically and radiographically (capping) in rickets. This excessive growth of the marginal regions through hyperplasia of the borderline chondroblast may result from the anabolic effect of PTH in this IGF-1-rich area. As rickets heals with vitamin D supplementation, some of these chondroblasts mature to become hypertrophic chondrocytes, and others differentiate to become sub-periosteal osteoblasts. Clinical widening of the wrist therefore remains evident for many years after rickets heals until it fades away through bone remodeling.

A gain-of-function mutation of the PTH/PTHrP receptor results in Jansen-type metaphyseal chondrodysplasia [70]. In this condition of constitutive activation of the receptor, irregular calcification and radiolucency are obvious **(figure 63)**. It is extreme in the radius and ulna, but the square form of the metacarpal epiphyses expresses the loss of normal spherical osteogenesis and the formation of parallel trabeculation.

Fig. 62
Rickets. Rachitic lower femur in an 8-month-old female who died. The very wide calcified cartilage and the less affected splayed marginal regions are evident of rickets. Photomicrograph. × 4.

Fig. 63
Jansen-type metaphyseal chondrodysplasia. In this condition of chronic activation of the receptor irregular calcification and radiolucency are obvious. It is extreme in the radius and ulna, but the square form of the metacarpal epiphyses expresses the loss of the normal spherical osteogenesis and the formation of parallel trabeculation.

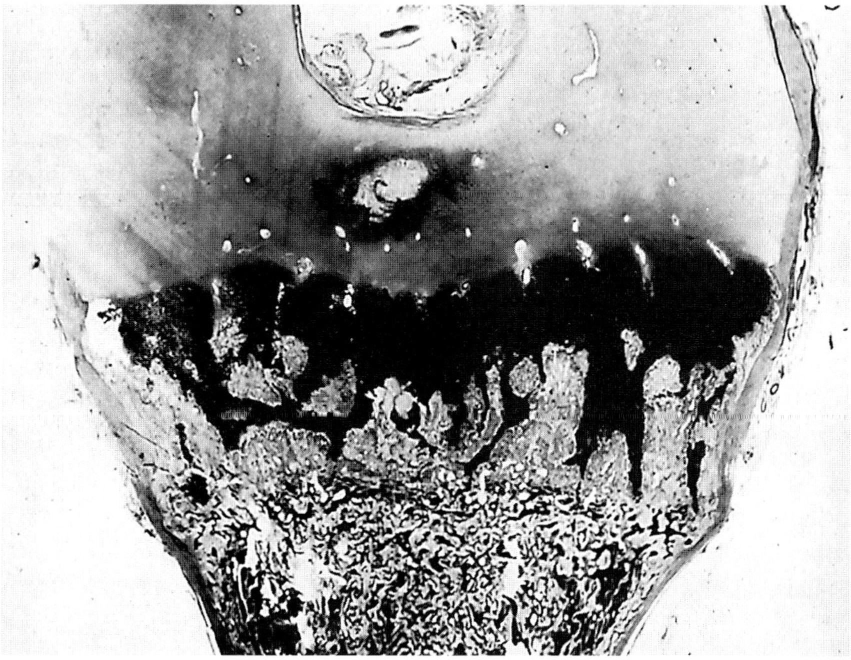

62

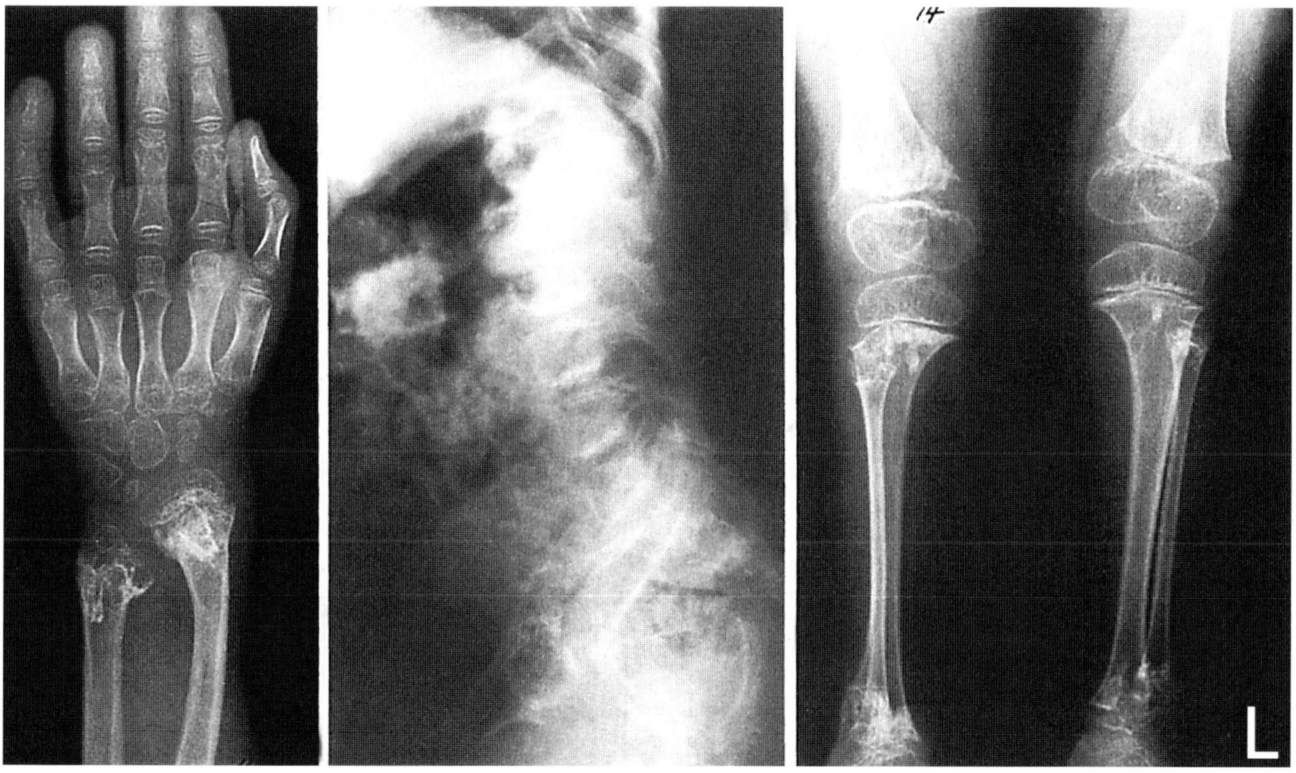

63

Glucocorticoids

Physiological Considerations

Long-term glucocorticoid excess causes growth retardation and osteoporosis. It does so through interference with osteoblast function, decreasing new bone formation. At the same time, the number of osteoclasts increase as does their binding to the bone surface [71]. Glucocorticoids downregulate IGF-1 generation by osteoblasts.

Indirect steroid-induced effects on bone include a decrease in renal calcium reabsorption and in intestinal calcium absorption and secondary increased plasma PTH levels. These occur despite normal levels of vitamin D and its metabolites and normal vitamin D receptor function. Other indirect effects include the glucocorticoid effects on muscle and connective tissue. The catabolic effect of glucocorticoids result in myopathy-related osteopenia and suppression of fibroblast DNA, RNA and protein.

Glucocorticoids suppress synthesis of both bone and cartilage extracellular matrix collagen and mineralization [72], whereas glycoprotein generation is enhanced. As a result of these changes in matrix function, osteogenesis is impaired, interfering with the normal transit from growth plate cartilage to the resorption front bone **(table 2)**.

Implications for Skeletal Maturity Readings

Glucocorticoid excess (Cushing syndrome) results from either Cushing disease or high-dose glucocorticoid therapy. The effects on cartilage and bone occur early in the course of the disease. The bone disease associated with glucocorticoid excess is bone atrophy rather than osteoporosis, seen in senescence or hyperparathyroidism – diminished formation rather than enhanced resorption.

Microscopy examination reveals sparse spongy trabeculae with few surrounding osteoblasts and thin cortex. Between the trabeculae, fat is more abundant than bone marrow is **(figure 64)**.

The deleterious effect of glucocorticoid excess on osteogenesis is apparent in bone maturation. In a study of 68 children with Cushing syndrome, long and short bones were affected minimally, whereas the carpal cuboid bones showed delayed maturation **(figure 65)**.

Fig. 64
Cushing's syndrome in a 30-year-old woman. Photomicrograph of the eleventh thoracic vertebra. Note the sparse spongy trabecules and the thin cortex. × 4.5.

Fig. 65
Cushing syndrome. The profile of bone maturation in 68 children and adolescents with Cushing syndrome, as compared to age-matched control group. Note that maximal delay is apparent in carpal bones.

Table 2. Glucocorticoid effects on cartilage and bone

Calcium metabolism	↓ Calcium intestinal absorption ↓ Calcium renal reabsorption ↑ Plasma PTH Normal vitamin D level and function
Cartilage	↓ IGF-1 generation ↓ Matrix collagen and mineralization ↑ Matrix proteoglycans
Bone	↓ Osteoblast function ↑ Osteoclast number and surface binding ↓ IGF-1 generation ↓ Matrix collagen and mineralization
Catabolic effects	Myopathy ↓ Fibroblast DNA, RNA and prot

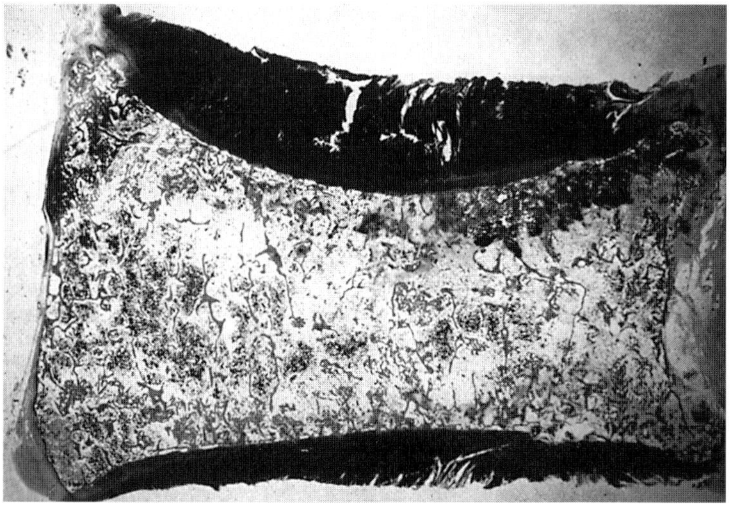

64

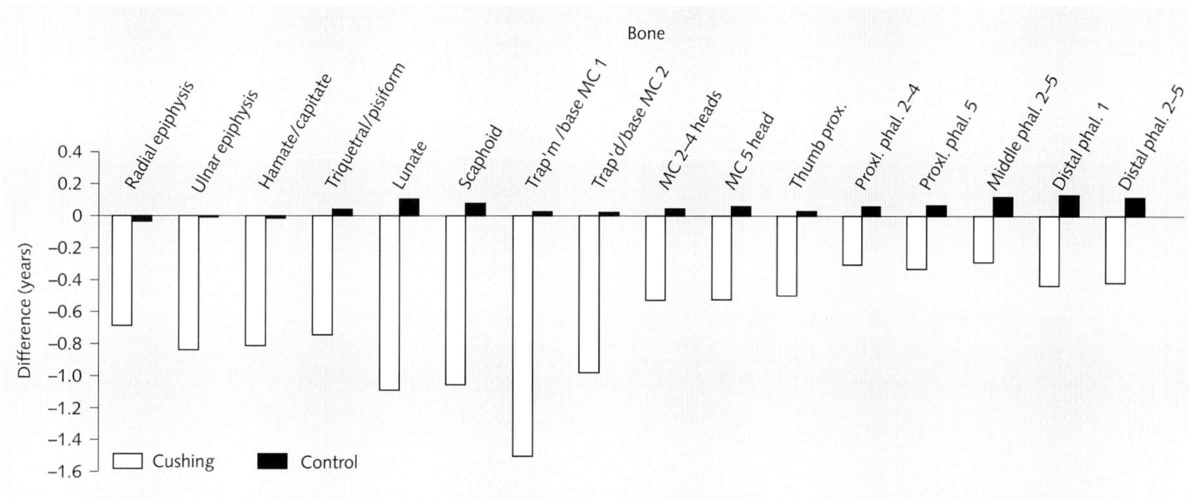

65

Other Diseases of Bone Growth and Maturation

Bone growth and maturation are affected in a large number of non-endocrine entities and their mechanisms are often unknown. Careful consideration of the hand and wrist films are important tools in the hands of the bone biologist in the clinical setting. The following is an incomplete list of such diseases.

Obesity

Patients with obesity have advanced bone maturation. **Figure 66** shows a radiogram of a child with morbid obesity and acanthosis nigricans at the age of 9 years and 7 months. Selective reading shows the differential maturation of his bones. Whereas his short bone maturation is advanced to an index 12, the carpal maturity is appropriate for age, at 9 years and 3 months, and the long bones maturity index is 10. This is a mirror image of bone maturation in Turner syndrome, and may implicate an effect of estrogen, as a product of intensive aromatization in adipose tissue.

Fig. 66
Obesity. Hand and wrist radiogram of 9 years and 7 months old boy with morbid obesity. Selective reading shows the differential maturation of his bones. Whereas his short bone maturation is advanced to an index 12, the carpal maturity index is appropriate for age, at 9 years and 3 months, and the long bones maturity index is 10.

Scurvy

Vitamin C is an essential cofactor in the generation of cartilage and bone matrix collagen. In its absence, matrix formation is inhibited, to be replaced by enhanced calcification. Scurvy results from vitamin C deficiency. The initial step of osteogenesis is affected. Cartilage matrix is replaced by excessive calcification ('white lines of scurvy') that does not allow for transition into bone, and the adjacent bone shows transverse translucency **(Figure 67)** [5]. The photomicrograph of the lower end of a femur of a subject with scurvy shows severe porosis of the secondary ossification center in the epiphysis. The hypertrophic zone shows excessive calcification and the adjacent metaphysis shows a fracture line that extends all the way across the width of the bone.

Fig. 67
Scurvy results from vitamin C deficiency.
a The initial step of osteogenesis is affected and the cartilage matrix is replaced by excessive calcification ('white lines of scurvy') that does not allow for transition into a bone. The adjacent bone shows transverse translucency.
b Lower end of a femur. Notice the severe porosis of the epiphyseal secondary ossification center. The hypertrophic zone shows excessive calcification and the adjacent metapysis shows a fracture line that extends all the way across the width of the bone. Photomicrograph. × 5.

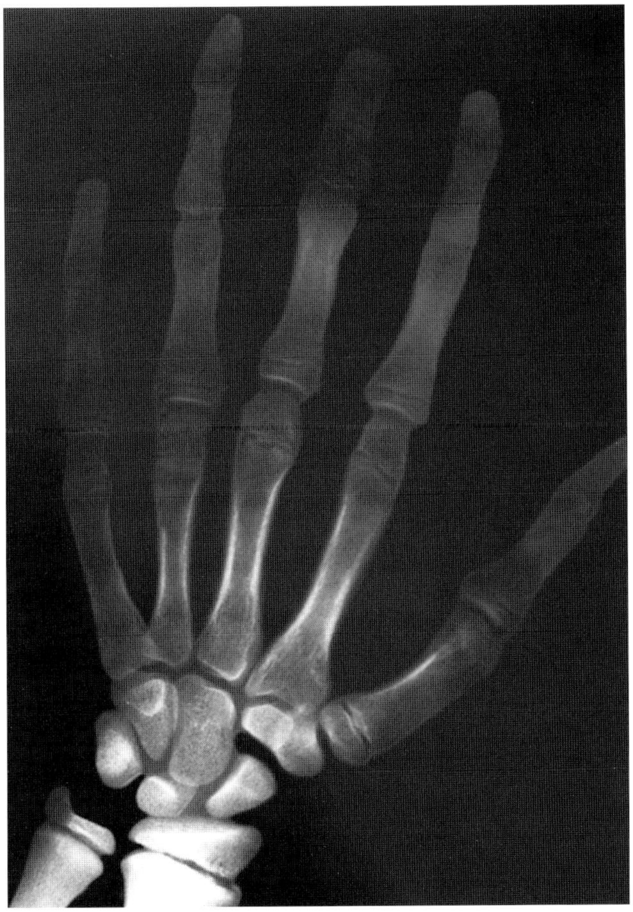

66

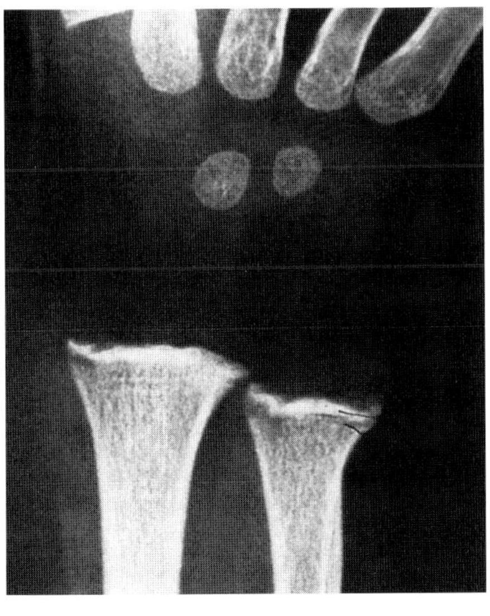

67a

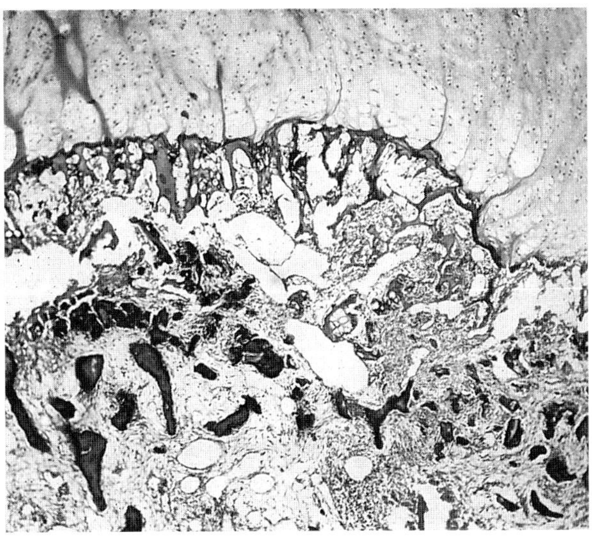

67b

Marfan Syndrome

Marfan syndrome results from an autosomally inherited mutation of the connective tissue protein fibrillin. The hand and wrist film **(Figure 68)** certainly gives the impression of spider's legs ('arachnodactyly') with their tapering digits. Pseudo-swelling around the joints arises from their lean soft tissue, but may also represent enhanced growth of the marginal growth plate regions. Can this be a mechanism for their tall stature? The slender shaft of the middle phalanges is reminiscent of hyperparathyroidism with enhanced bone resorption. Bone maturity index is uneven in this 14-year-old boy. Short bones index is 10.5, long bones index 12 and carpal bone index 13. These children with excessive final height often apply for growth-slowing therapy. Due to these dramatic effects on bone maturation, height prediction, based on maturation indices, may be inaccurate when using any of the available methods.

Fig. 68
Marfan syndrome. The hand and wrist film gives the impression of spider's legs ('arachnodactyly') with their tapering digits. Pseudo-swelling around the joints arises from their slender shafts of the middle phalanges that are reminiscent of hyperparathyroidism.

Renal Failure

Renal failure results in severe growth arrest arising from many possible mechanisms. The metabolic milieu of renal failure with chronic acidosis, disturbed calcium metabolism and secondary hyperparathyroidism are but some of those mechanisms. Anorexia, malnutrition and the catabolic state that ensue are additional factors that are associated with decreased serum levels of IGF-1. **Figure 69** shows the hand radiogram of a 28-month-old female with renal failure, while she was on daily peritoneal dialysis. Her height and weight were – 2.3 SDS for age. Skeletal maturity has an index of 1 year and 4 months, but the information retrieved from the film is beyond the 'bone age'. The resorption front, where cartilage transforms into bone, is seen as a pseudo-growth plate in both the radius and ulna, suggesting dissociation across that interface, where osteogenesis takes place. The thin cortex may be related to steroid therapy or to secondary hyperparathyroidism.

Fig. 69
Renal failure. Radiogram of a 28-month-old girl with renal failure, while she was treated with daily peritoneal dialysis. Her height and weight were – 2 to 3 SDS for age. Skeletal maturity of long, short and cuboid bones has an index of 14/12. The resorption front, where cartilage transforms into bone, is seen as a pseudogrowth plate in both the radius and ulna (arrows), suggesting dissociation across that interface. The thin cortex may be related to steroid therapy or to secondary hyperparathyroidism.

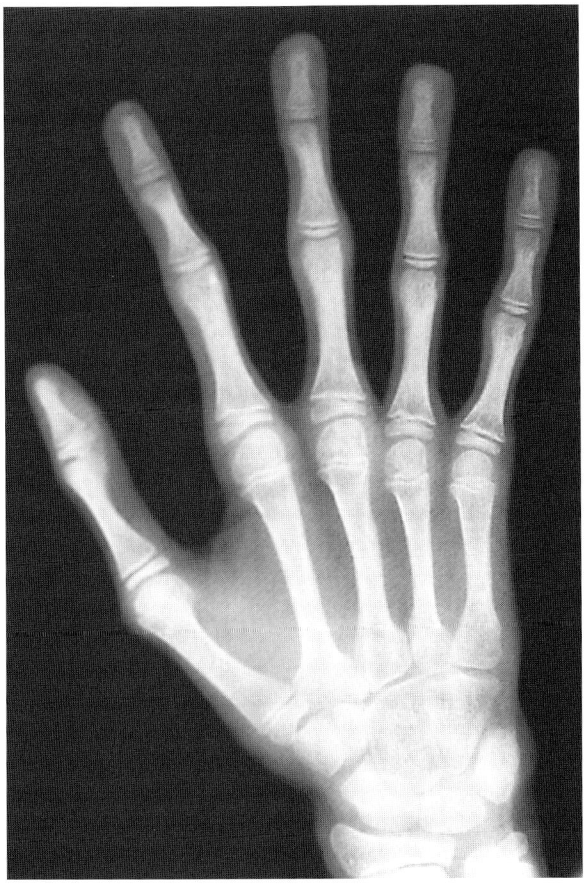

68

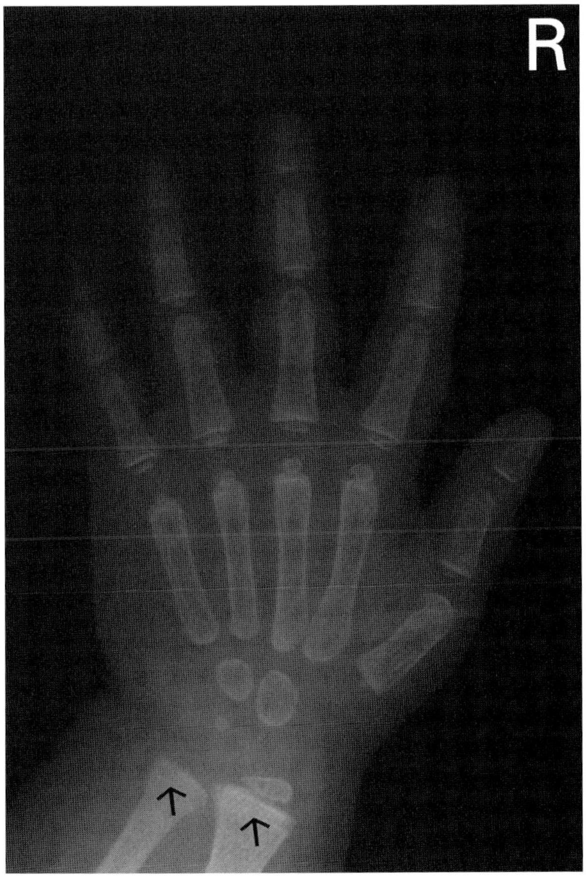

69

Assessment of Skeletal Maturity– Critical Comments

From the very beginning of this discipline, human anthropologists have appreciated the importance of longitudinal follow-up of skeletal maturity. The earliest attempt to develop a standard of reference was reported in 1937 in Todd's Atlas of Skeletal Maturation [3]. Dental age has been advocated, mostly in Scandinavia, and the methodology is well developed [73], but the variation is so great that it provides little help in clinical practice.

This monograph does not attempt to review or be critical of all the published methods, but rather to mention the two most widely used. It does not strive to instruct the reader how to use these methods, but rather, to give an overview and be critical of the two available concepts: the atlas method and the individual bones system. Owing to the different concept of these two methods, they are not always comparable and cannot be considered interchangeable. With the two-dimensional representation of the hand, a slight bend or rotation in the hand position gives an ambiguous appearance and increases the inter- and intraobserver variability.

Greulich and Pyle Atlas Method

Developed from films taken in the 1930s and 1940s, this is probably still the most frequently used method [74]. The atlas calls for a comprehensive analysis of bones of the hand and wrist, compared to a standard set from the median of a range evaluated for each chronological age. The technique of maturity reading varies among raters. Greulich and Pyle themselves suggested that raters should adapt to individual needs or preferences [74]. Although instructions are given for individual assessment of each bone and epiphysis [75], the user tends to be influenced by the rest of the hand. Despite the clear instructions, it has been implied that the original concept of the atlas method was that skeletal maturation is even [76]. The series used in the original atlas is still in use today and it is obvious to experienced raters that they do not represent an ideal set of radiographs and that in many of the original radiographs various stages of bone maturity were uneven. Nevertheless, a recent reevaluation of each individual bone in a cohort of normal individuals showed the reliability of the system to be accurate with deviations not greater than 6 months **(figure 65)**.

TW2 Individual Bones Method

Scoring bone by bone [77], this system is potentially superior to the atlas method. The longitudinal nature of the series of 113 radiographs and the scoring system accommodates for the fact that the data were collected over a period of 20 years as of the 1950s. Moreover, the scoring of individual bones allows for a more meaningful statistical analysis when searching for the biological meaning of each bone's maturation. Yet, as soon as each bone is assessed, the assessor is asked to desert all this useful information from individual bones and give a combined score that gives more weight to the long bones and the radius and ulna than to the short bones. The scoring system even accounts for sex differences and thereby eliminates the very clear and unique effects of sex steroids. Then the user must abandon the rationale of the scoring system to give a unified 'bone age'.

The excessive weight given to the long bones entails a potential for apparent discontinuous maturation. Thus, the progression of the radius and ulna scores from stages F to G, with no intermediate stage, increases the maturation index from 11 to 12.

The TW2 method intermingles measures of bone maturity with size. The artificial attempt to grade each bone into 8 or 9 maturity stages gives extra weight to minor insignificant changes. Poor positioning of the hand when the radiograph is taken alters the radiographic appearance of the epiphyses and makes interpretation difficult and inconsistent [78].

The use of computed image analysis of bone maturation might have increased the precision and reproducibility of reading. In its newest version [79] it allows for interpolation between stages, to produce a continuous score, that is, however, still an average of several bones, losing their singular identity.

Adult Height Prediction

Skeletal maturity index correlates in normal individuals closely with the remaining growth potential, and as such has been used to predict the fusion of growth plates. This tool has been used in predicting adult height in normal children, in short or tall children, as well as in patients with growth disorders. It is essential to perceive that delayed bone maturity does not necessarily reflect increased growth potential, but can rather resonate a chronic disease with compromised potential. Thus, children with hypothyroidism or GH deficiency have delayed bone maturation but do not have a better potential for later growth. On the other hand, patients with aromate deficiency or defective estrogen receptors keep growing despite their retarded bone age.

In studies on the accuracy of predictions, the mean values were acceptable at ± 1 cm, but the distribution of predictions is as wide and unacceptable as ± 6 cm (95% confidence limits). The reasons lie with the precision of assessments and with the unified 'bone age', but also with our limited knowledge of the bones that need to be emphasized. We do not know whether long or short bone maturity best represents that of the leg bones.

Skeletal Maturity as an Argument in Functions

The terminology of 'bone age' and its assignment of 'years' as its units are problematic. Age and its unit years are precise measures with no variance, and form an ideal argument in a function. The use of bone age 'years' (or index) as an argument in a function, such as in growth charts, brings in a great deal of imprecision. It has become a common practice to plot auxological data or even laboratory results against bone age 'years'. This is entirely unjustified. First of all, no normative data were ever collected against an argument of skeletal maturation. The assumption that a bone maturity index can replace the chronological age on the x-axis has not been tested scientifically. In fact, in many clinical conditions it does not [2] as skeletal maturation lags many months behind abrupt endocrine changes such as onset of normal or abnormal puberty or the outset of a disease.

The 'age' in 'bone age' insinuates a concept of precision and unity, which it does not provide. This is due to individual variability, imprecise quasi-quantification and the individual patterns of bone maturation. As mentioned above, both inter- and intraobserver variability may be extremely high. This is created by the long experience it takes to master the art of hand and wrist radiology, but also by the disposition towards various bones in the film and the decision to use average, sum scores or specific bones only, such as the RUS method.

Deviations of the skeletal maturity of children from their chronological ages are as common as deviations of any biological parameter (height, serum thyroxine, etc.) from its mean value. The distribution is narrow in infancy and increases gradually during childhood to maximal deviation at near-puberty. In the study of growth, we have accepted the practice of expressing stature as 'standard deviation score' (SDS). This has freed us from the misconcept of 'height age' or 'weight age', entailing an optimal growth at the 50th percentile. It would be sensible to use SDS values in expressing skeletal maturity; yet, such normative values are still to be developed.

Practical Guidelines

There is more than one way to read bone maturity films. The following are personal steps I take in reading them.

- I use the Greulich and Pyle Atlas.
- To refrain from a bias by the child's age, I do not look at the plates' title up to the very end.
- I start by making a rough estimate of the film. In young children I use primary and secondary ossification centers for that purpose. In late adolescence, I use epiphyseal fusion and in between I use the maturity indicators of the Atlas.
- I read bone groups following a regular order. I read long bones (radius and ulna) first, then the carpal bones and finally the short bones (metacarpals and phalanges).
- Each bone group attains its own maturity index. Missing ossification centers are labeled 'Less than …'. It is important to realize that unified maturity is uncommon. Do not average the scores as you lose the perception of bone biology in the clinical setting.
- Each bone group is then compared to the next and previous plates. If you feel experienced, you can give intermediate scores.
- I then try to identify the outlined most retarded bones. I scrutinize for the ossification age of the respective bones and discuss possible insults during that age with patient and family.
- The film is also viewed as diagnostic radiography. Look for disease-specific pathology, growth arrest lines, irregular epiphyses, etc.

- The film is compared to previous bone maturity films of the same subject. The change (Δ) in bone maturity is related to the Δ chronological age for each of the bone groups (Δ maturity score/Δ chronological age). This ratio is then related to the growth velocity.
- Now that all this information is available, one can contemplate bone biology and mechanism of disease.

Osteoporosis

It is generally felt that X-ray films have limited value in the assessment of bone density and osteoporosis, other than detecting fractures. However, the routinely used films of the wrist and hand do have the capacity to provide rough estimates of cortical bone thickness.

Based on a method developed in the 1960s [80] there is a recent new interest in metacarpal cortical thickness in the assessment of bone density. Measuring cortical thickness in the second metacarpal bone is a valuable tool in the hands of an experienced observer. This was later developed into computed radiography [81], but the principle remains the same. The 'metacarpal index' is the combined cortical thickness (both sides) normalized with regard to the outer bone diameter of the measuring site, the midshaft of the second metacarpal, or the three midmetacarpals of both hands. It is reduced with age, particularly in postmenopausal women, and correlates with axial bone mass in group studies. Measurement of the cortical thickness in its modern version, digital X-ray radiogrammetry, requires only a plain analog radiograph, a PC, a film scanner, and reliable software. It can be used diagnostically and longitudinally for monitoring changes. Cortical thickness can be measured on old radiographs, thus enabling estimation of cortical bone loss or gain from the time of earlier recordings [82] **(figure 70)**.

Fig. 70
Cortical thickness: The metacarpal index is the combined cortical thickness (both sides) normalized with regard to outer bone diameter (arrows) of the measuring site at the midshaft of the second metacarpal.
a The metacarpal index of this 21-month-old female is normal, at 0.28.
b The metacarpal index of this 12-year-old boy is 0.58.

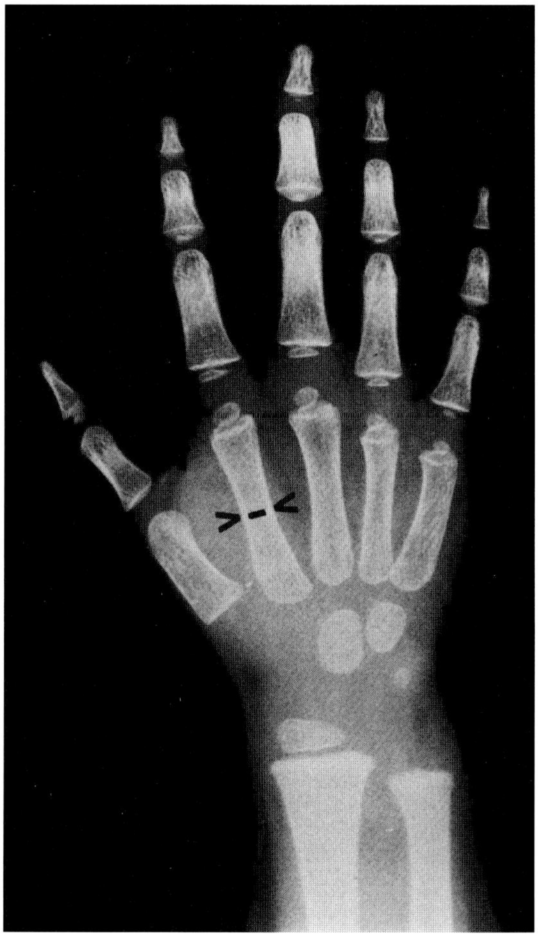

70a

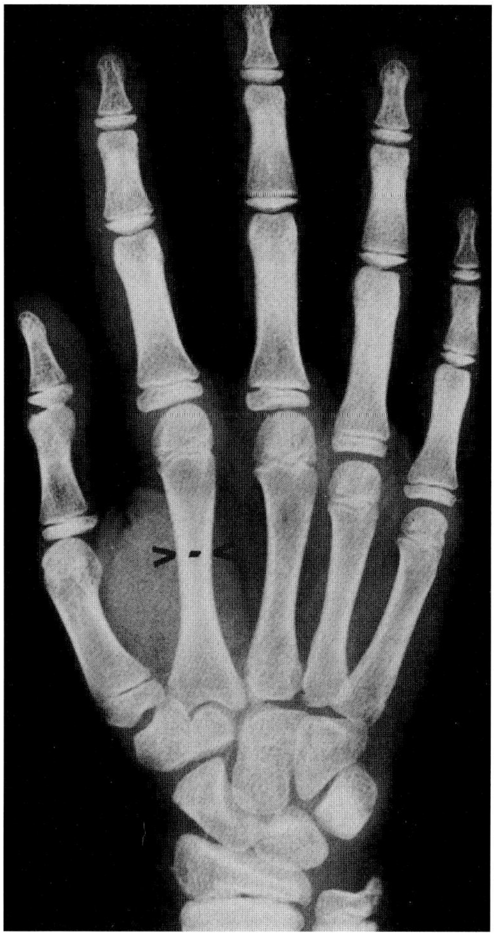

70b

Conclusions

The principal message is that the analysis of a bone maturation film provides the clinician with a lot more than only a 'bone age' reading. It provides a unique opportunity to glimpse into the biology of bone, to contemplate mechanisms of growth and to consider the hormones and growth factors that contribute to bone growth and development. As such, these X-ray films are powerful research tools in the hands of the clinician. Comparing the hand radiogram in a given diagnosis to the known histology, human or experimental pathology of bones in that same condition renders a strong relationship between X-ray and mechanism. This can then be used to analyze unknown clinical conditions.

To take advantage of the entire information provided by a hand and wrist film, one must keep in mind that different bones undergo different developmental processes. In the simplest way, long bones grow differently than do cuboid or flat bones. Epiphyses grow in the bone's longitudinal axis (in both directions) but also sideways. Developmentally, each axis often has its own unique growth factors and critical time periods. The center of a growth plate has a different growth mechanism than does the marginal region of that same plate. Finally, miscellaneous bones have different mechanisms. In the most elemental way, the radius and the metacarpals, both being long bones progress often in a different tempo. Because they do so, I call the metacarpals and phalanges 'short bones'. The message is to read each bone separately, to document the development of each bone separately and to not average those readings.

Even if the maturity index designation takes time to be adopted, the labeling of bone maturity units as 'years' has to remain within apostrophes. Whereas the calendar unit year is a precise measure, bone maturity reading is subjective. The use of 'years' units often deludes patients to apprehend bone age 'years' as a means to evaluate the child's biological age. I prefer the use of the same measures as index units.

Whereas this monograph is meant to broaden the utilization of the hand and wrist X-ray films, the bone biologist in the clinical setting has other bedside tools. Body proportions give important information on the growth of unique body parts. Magnetic resonance imaging gives a powerful point of view at the growth plate. Ultrasonograpy of the growth plate is a relatively new modality.

Basic research enhances our understanding of the molecular and structural mechanisms which pertain to bone growth and maturation. The clinical aspects of these developments are in the hands of the bone biologist in the clinical setting.

References

1 Greulich WW, Pyle SI: Radiographic Atlas of the Hand and Wrist. Stanford, Stanford University Press, 1959.

2 Benso L, Vannelli S, Pastorin L, Benso A, Milani S: Variation of bone age progression in healthy children. Acta Paediatr Suppl 1997;23:109 – 112.

3 Todd TW: Atlas of Skeletal Maturation (Hand). St Louis, Mosby, 1937.

4 Jaffe HL: Metabolic, Degenerative and Inflammatory Diseases of Bones and Joints. Philadelphia, Lea & Febiger, 1972.

5 Caffey J: Pediatric X-ray diagnosis, ed 7. Chicago, Year Book Medical Publishers, 1978.

6 Ganey TM, Love SM, Ogden JA: Development of vasculature in the chondroepyphisis of the rabbit. J Orthop Res 1992;10: 496 – 510.

7 Hunziker EB, Wagner J, Zapf J: Differential effects of IGF-I and hGH on the various developmental stages of growth plate chondrocytes in vivo. J Clin Invest 1994;3:1078 – 1083.

8 Hunziker EB, Schenk RK, Cruz Orive LM: Quantitation of chondrocyte performance in growth-plate cartilage during longitudinal bone growth. J Bone Joint Surg (Am) 1987;9:162 – 173.

9 Hunziker EB: Mechanism of longitudinal bone growth and its regulation by growth plate chondrocytes. Micros Res Techn 1994;8:505.

10 Wilsman NJ, Farnum CE, Leiferman EM, Fry M, Barreto C: Differential growth by growth plates as a function of multiple parameters of chondrocytic kinetics. J Orthop Res 1996;14: 927 – 936.

11 Gerber H-P, Vu TH, Ryan AM, Werb Z, Ferrara N: VEGF couples hypertrophic cartilage remodeling, ossification and angiogenesis during enchondral bone formation. Nat Med 1999;5: 623 – 628.

12 Roach Hi, Erenpreisa J, Aigner T: Osteogenic differentiation of hypertrophic chondrocytes involves asymmetric cell divisions and apoptosis. J Cell Biol 1995;31:483 – 494.

13 Gibson G: Active role of chondrocyte apoptosis in enchondral ossification. Microsc Res Techn 1998;3:191 – 204.

14 Gray's Anatomy: Edinburgh, Churchill-Livingstone, 1995.

15 Pavasant P, Shizari T, Underhill CB: Hyaluronan contributes to the enlargement of hypertrophic lacunae in the growth plate. J Cell Sci 1996;9:327 – 334.

16 Lee ER, Lumplugh L, Shepard NL, Mort JS: The septoclast, a cathepsin B-rich cell involved in the resorption of growth plate cartilage. J Histochem Cytochem 1995;3:525 – 536.
17 Lewinson D, Shenzer P, Hochberg Z: Growth hormone involvement in the regulation of tartrate-resistant acid phosphatase-positive cells that are active in cartilage and bone resorption. Calcif Tissue Int 1993;2:216 – 222.
18 Robertsom WW Jr: Newest knowledge of the growth plate. Clin Orthop Related Res 1990;253:270 – 278.
19 Miralles-Flores C, Delgado-Baeza E: Histomorphometric differences between the lateral region and central region of the growth plate in fifteen-day-old rats. Acta Anat 1990;39:209 – 213.
20 Lazowski DA, Fraher LJ, Hodsman A, Steer B, Modrowski D, Han VKM: Regional variation of insulin-like growth factor-I gene expression in mature rat bone and cartilage. Bone 1994;5:563 – 576.
21 Tanner JM, Whitehouse RH, Cameron N, Marshall WA, Healy MJR, Goldstein H: Assessment of Skeletal Maturity and Prediction of Adult Height (TW-2 Method), ed 2. London, Academic Press, 1983.
22 Edeiken J, Hodes PH: Roentgen Diagnosis of Diseases of Bone. Baltimore, Williams & Wilkins, 1973.
23 O'Rahilly R: A survey of carpal and tarsal anomalies. J Bone Joint Surg (Am) 1953;5:626.
24 Ecklund K, Doria AS, Jaramillo D: Rickets on MR images. Pediatr Radiol 1999;9:673 – 675.
25 Dott NM, Fraser J: The influence of experimental pituitary and thyroid derangements upon the developmental growth of bone. J Exp Physiol 1923;3:107 – 120.
26 Barnard R, Haynes KM, Werther GA, Waters MJ: The ontogeny of growth hormone receptors in the rabbit tibia. Endocrinology 1988;22:2562 – 2569.
27 Maor G, Hochberg Z, v d Mark K, Heinegard D, Silbermann M: Human growth hormone enhances chondrogenesis and osteogenesis in a tissue culture system of chondroprogenitor cells. Endocrinology 1989;25:1239 – 1245.
28 Maor G, Hochberg Z, Silbermann M: Insulin like growth factor-I accelerates proliferation and differentiation of cartilage progenitor cells in cultures of neonatal mandibular condyles. Acta Endocrinol 1993;28:56 – 64.

29 Ohlsson C, Nilsson A, Isaksson OGP, et al: Growth hormone induces multiplication of the slowly cycling germinal cells of the rat tibial growth plate. Proc Natl Acad Sci USA 1992;9:9826–9831.
30 Trippel SB, Wroblewski J, Makower A-M, Whelan MC, Schoenfeld D, Doctrow SR: Regulation of growth-plate chondrocytes by insulin-like growth factor-I and basic fibroblast growth factor. J Bone Joint Surg 1993;5A:177–189.
31 Ohlsson C, Bengtsson B-A, Isaksson OGP, Andreassen TT, Slootweg MC: Growth hormone and bone. Endocr Rev 1998;9: 55–79.
32 Hou P, Sato T, Hoffstetter W, Foged NT: Identification and characterization of the insulin-like growth factor I receptor in mature rabbit osteoclasts. J Bone Miner Res 1997;2:534–540.
33 Hill PA, Reynolds JJ, Meikle MC: Osteoblasts mediate insulin-like growth factor I and -II stimulation of osteoclast formation and function. Endocrinology 1995;36:124–131.
34 Boersma B, Wit JM: Catch-up growth. Endocr Rev 1997;18:646–661.
35 Scow RO, Simpson ME, Asling CW, Li CH, Evans HM: Response by the rat thyro-parathyroidectomized at birth to growth hormone and to thyroxine given separately and in combination. Anat Rec 1949;104:445–454.
36 Lewinson D, Harel Z, Shenzer P, Silbermann M, Hochberg Z: Effect of thyroid hormone and growth hormone on recovery from hypothyroidism of epiphyseal growth plate cartilage and in its adjacent bone. Endocrinology 1989;124:937–945.
37 Allain TG, McGregor AM: Thyroid hormones and bone. J Endocrinol 1993;139:9–18.
38 Franklin J, Betteridge J, Holder R, Daykin J, Lilley J, Sheppard M: Bone mineral density in thyroxine treated females with or without a previous history of thyrotoxicosis. Clin Endocrinol 1994;41:425–432.
39 Williams GR, Robson H, Shalet SM: Thyroid hormone actions on cartilage and bone: interaction with other hormones at the epiphyseal plate and effects on linear growth. J Endocrinol 1998;157:391–403.
40 Yaffe BM, Samuels HH: Hormonal regulation of growth hormone gene: Relationship of the rate of transcription to the level of nuclear thyroid hormone receptor complexes. J Biol Chem 1984;259:6284–6290.

41 Hochberg Z, Bick T, Harel Z: Alterations of human growth hormone binding by rat liver membranes during hypo- and hyperthyroidism. Endocrinology 1990;26:325 – 329.
42 Lewinson D, Harel Z, Shenzer P, Silbermann M, Hochberg Z: Effect of thyroid hormone and growth hormone on recovery from hypothyroidism of epiphyseal growth plate cartilage and its adjacent bone. Endocrinology 1989;24:937 – 945.
43 Burch WM, Van Wyk JJ: Triiodothyronine stimulates cartilage growth and maturation by different mechanisms. Am J Physiol 1987;252:E176 – E183.
44 Alini M, Kofski Y, Wu W, Pidoux I, Poole AR: In serum-free culture thyroid hormones can induce full expression of chondrocyte hypertrophy leading to matrix calcification. J Bone Miner Res 1996;11:105 – 113.
45 Meema HE, Schatz DL: Simple radiologic demonstration of cortical bone loss in thyrotoxicosis. Radiology 1970;97:9 – 15.
46 Braidman IP, Daveport LK, Carter DH, Selby PL, Mawer EB, Freemont AJ: Preliminary in situ identification of estrogen target cells in bone. J Bone Miner Res 1995;10:74 – 80.
47 Manolagas SC, Jilka RL: Bone marrow, cytokines and bone remodelling. N Engl J Med 1995;332:305 – 311.
48 Hofbauer LC, Khosla S: Androgen effects on bone metabolism: Recent progress and controversies. Eur J Endocrinol 1999;140: 271 – 286.
49 Carrascosa A, Audi L, Ferrandez MA, Ballabriga A: Biological effects of androgens and identification of specific dihydrotestosterone binding sites in cultured human fetal epiphyseal chondrocytes. J Clin Endocrinol Metab 1990;70:134 – 140.
50 Somjen D, Weisman Y, Mor Z, Harell A, Kaye AM: Regulation of proliferation of rat cartilage and bone by sex steroid hormones. J Steroid Biochem Molec Biol 1991;40:717 – 723.
51 Mauras N, Haymond MW, Vienna NE, Abrams SA, Yergey AL: Calcium and protein kinetics in prepubertal boys: Positive effects of testosterone. J Clin Invest 1994;3:1014 – 1019.
52 Wierman ME, Beardsworth DE, Crawford JD, Crigler JF Jr, Mansfield MJ, Bode HH, Boepple PA, Kushner CD, Crwley WF Jr: Adrenarche and skeletal maturation during luteinizing hormone releasing hormone analogue suppression of gonadarche. J Clin Invest 1986;77:121 – 126.

53 Smith EP, Boyd J, Frank GR, Takahashi H, Cohen RM, Specker B, Williams TC, Lubahn DB, Korach KS: Estrogen resistance caused by a mutation in the estrogen-receptor gene in man. N Engl J Med 1994;31:1056 – 1061.
54 Morishima A, Grumbach MM, Simpson ER, Fisher C, Qin K: Aromatase deficiency in male and female siblings caused by a novel mutation and the physiological role of estrogens. J Clin Endocrinol Metab 1995;80:3689 – 3698.
55 Finby N, Archibald RM: Skeletal abnormalities associated with gonadal dysgenesis. Am J Radiol 1965;93:354 – 361.
56 Mehta AV, Chidambaram B, Suchedina AA, Garrett AB: Radiological abnormalities in Turner's syndrome. Chest 1993;104: 1795 – 1799.
57 Webber ML, Puck MH, Maresh MM, Goad WB, Robinson A: Skeletal maturation of children with sex chromosome abnormalities. Pediatr Res 1982;16:343 – 346.
58 Acheson RM, Zampa GA: Skeletal maturation in ovarian dysgenesis and Turner's syndrome. Lancet 1961;i:917 – 919.
59 Park E: Radiological anthropometry of the hand in Turner's syndrome. Am J Phys Anthropol 1977;46:463 – 470.
60 Even L, Bronstein V, Hochberg Z: Bone maturation in girls with Turner's syndrome. Eur J Endocrinol 1998;138:59 – 62.
61 Kato Y, Shimazu A, Iwamoto M, Nakashima K, Koike T, Suzuki F, Nishii Y, Sato K: Role of 1,25-dihydroxycholecalciferol in growth-plate cartilage: inhibition of terminal differentiation of chondrocytes in vitro and in vivo. Proc Natl Acad Sci USA 1990; 87:6522 – 6526.
62 Inao A, Conrad HE: Coordinate inhibition of alkaline phosphatase and type X collagen synthesis by 1,25-dihydroxyvitamin D_3 in primary cultured hypertrophic chondrocytes. Calcif Tissue Int 1992;50:445 – 450.
63 Hochberg Z, Tiosano D, Even L: Calcium therapy in calcitriol resistant rickets. J Pediatr 1992;21:803 – 808.
64 Schwartz Z, Semba S, Graves D, Dean DD, Sylvia VL, Boyan BD: Rapid and long-term effects of PTH(1 – 34) on growth plate chondrocytes are mediated through two different pathways in a cell maturation-dependent manner. Bone 1997;21:244 – 259.
65 Pines M, Hurwitz S: The effect of parathyroid hormone and atrial natriuretic peptide on cyclic nucleotides production and proliferation of avian epiphyseal growth plate chondroprogenitor cells. Endocrinology 1988;123:360 – 365.

66 Rihani-Bisharat S, Maor G, Lewinson D: In vivo anabolic effects of parathyroid hormones (PTH)28 – 48 and N-terminal fragments of PTH and PTH related protein on neonatal mouse bones. Endocrinology 1998;139:974 – 981.
67 Lee K, Deeds JD, Segre GV: Expression of parathyroid hormone-related peptide and its receptor messenger ribonucleic acids during fetal development of rats. Endocrinology 1995;136:453 – 463.
68 Amling M, Neff L, Tanaka S, Inoue D, Kuida K, Weir E, Philbrick WM, Broadus AE, Baron R: Bcl-2 lies downstream of parathyroid hormone-related peptide in a signaling pathway that regulates chondrocyte maturation during skeletal development. J Cell Biol 1997;36:205 – 213.
69 Hunter WL, Arsenault AL, Hodsman AB: Rearrangement of the metaphyseal vasculature of the rat growth plate in rickets and rachitic reversal: A model of vascular arrest and angiogenesis renewed. Anat Rec 1991;29:453 – 461.
70 Schipani E, Kruse K, Juppner H: Constitutively active mutant PTH-PTHrP receptor in Jansen-type metaphiseal chondrodysplasia. Science 1995;268:98 – 100.
71 Hahn TJ, Halstead LT, Teitelbaum SL, et al: Altered mineral metabolism in glucocorticoid induced osteopenia: effect of 25-hydroxyvitamin D administration. J Clin Invest 1979;4: 655 – 665.
72 Silberman M, Maor G: Mechanism of glucocorticoid induced growth retardation: Impairment of cartilage mineralization. Acta Anat 1978;101:140 – 149.
73 Demirjian A, Goldstein H, Tanner JM: A new system of dental age assessment. Hum Biol 1973;5:11 – 27.
74 Gruelich WS, Pyle SI: Radiographic Atlas of Skeletal Development of the Hand and Wrist. Stanford, Stanford University Press, 1959.
75 Pyle SI, Reed RB, Stuart HC: Patterns of skeletal development in the hand. Pediatrics 1959;4:886 – 903.
76 Gilli G: The assessment of skeletal maturation. Horm Res 1996; 5(suppl 2):49 – 52.
77 Tanner JM, Whitehouse RH, Marshall WA, et al: Assessment of Skeletal Maturity and Prediction of Adult Height: TW2 Method. New York, Academic Press, 1975.
78 Cox LA: Tanner Whitehouse method of assessing skeletal maturity: Problems and common errors. Horm Res 1996;5(suppl 2): 53 – 55.

79 Tanner JM, Gibbons RD: Automatic bone age measurements using computerized image analysis. J Pediatr Endocrinol 1994;7:141 – 145.
80 Spencer RP, Garn SM, Coulombe MJ: Age-dependent changes in metacarpal cortical thickness in two populations. Invest Radiol 1966;1:394 – 397.
81 Rico H, Revilla M, Fraile E, Martin FJ, Cardenas JL, Villa LF: Metacarpal cortical thickness by computed radiography in osteoporosis. Bone 1994;15:303 – 306.
82 Nielsen SP: The Metacarpal Index revisited: A brief overview. J Clin Densitom 2001;4:199 – 207.

Photographic Credits

Fig. 3, 5b, 13, 14, 19, 20, 25, 29, 38, 39, 41, 56, 59, 62, 64 from Jaffe [4].

Fig. 6, 66 from Caffey [5].

Fig. 8 from Hunziker [9].

Fig. 10, 11, 21 from [14].

Fig. 12 from Lewinson et al. [17].

Fig. 15 from Robertsom [18].

Fig. 16 from Miralles-Flores and Delgado-Baeza [19].

Fig. 18, courtesy D. Lewinson, Haifa.

Fig. 27 from Edeiken and Hodes [22].

Fig. 28 from O'Rahilly [23].

Fig. 32, courtesy V. Bialik, Haifa.

Fig. 33 from Ecklund et al. [24].

Fig. 34, courtesy G. Maor, Haifa.

Fig. 36, courtesy Z. Laron, Tel-Aviv.

Fig. 37 from Lewinson et al. [36].

Fig. 44, 45 from Meema and Schatz [45].

Fig. 52, courtesy E. Smith, Cincinnati, Ohio.

Fig. 55 from Even et al. [60].

Fig. 61 from Amling et al. [68].

Fig. 63 from Schipani et al. [70].

Fig. 65, courtesy T. Lafferty, Perth.

Subject Index